Structural Failure in Residential Buildings

Volume 3   Basements and Adjoining Land Drainage

The publishers are grateful for the help and advice of John G. Roberts B.Arch., FRIBA, of the Welsh School of Architecture, University of Wales Institute of Science and Technology, in the preparation of the English language edition of this book.

Structural Failure in Residential Buildings

Volume 3
Basements and Adjoining Land Drainage

Erich Schild
Rainer Oswald
Dietmar Rogier
Hans Schweikert

Illustrations by Volker Schnapauff

GRANADA
London Toronto Sydney New York

Granada Publishing Limited – Technical Books Division
Frogmore, St. Albans, Herts AL2 2NF
and
3 Upper James Street, London W1R 4BP
866 United Nations Plaza, New York, NY 10017, USA
117 York Street, Sydney, NSW 2000, Australia
100 Skyway Avenue, Rexdale, Ontario M9W 3A6 Canada
PO Box 84165, Greenside, 2034 Johannesburg, South Africa
CML Centre, Queen & Wyndham, Auckland 1, New Zealand

Copyright © Granada Publishing, 1980

Translated from the German by TST Translations

ISBN 0 246 11170 4

First published in 1978 in the Federal Republic of Germany
by Bauverlag GmbH, Wiesbaden and Berlin
English edition first published in Great Britain 1980
by Granada Publishing – Technical Books Division

Printed in Great Britain by William Clowes (Beccles) Limited
Beccles and London

Granada®
Granada Publishing ®

British Library Cataloguing in Publication Data

Structural failure in residential buildings.
   Vol. 3: Basements and adjoining land drainage
   1. Dwellings – Defects
   I. Schild, Erich II. Schnapauff, Volker
   III. Roberts, John G
   690'.8   TH4812
   ISBN 0-246-11170-4

This publication is the result of the research project 'Problems of structural failure –
prevention of structural failure in residential buildings', commissioned by the
Ministry of the Interior (Residential Buildings Department) of North Rhein-
Westphalia and carried out at the Technical University of Aachen, Faculty of Building
Construction III – Building Science and Problems of Structural Failure (senior
professor Prof. Dr.-Ing. Erich Schild).
Research team:

Erich Schild	Prof. Dr.-Ing.
Rainer Oswald	Dipl.-Ing.
Dietmar Rogier	Dipl.-Ing.
Volker Schnapauff	Dipl.-Ing.
Hans Schweikert	Dipl.-Ing.
Detlef Bock	cand. arch.
Elisabeth Teck	cand. arch.
Jürgen Roder	cand. arch.

Norma Gottstein
Gunda Hoppe

Preface to English edition

Developments resulting from the extensive investigations into the performance of building elements above ground level reflect a better understanding of the principles governing their behaviour. The same cannot be said for such elements below ground; basement construction has been neglected and still presents problems.

The conventional basement in British houses was generally used for purposes of utility, storage and heating. Consequently their internal environment and structural properties have been inadequately considered.

With the development of hill-side housing, the decline in high-rise building and the rising costs of sites, increasing use is being made of basement accommodation for normal room purposes. To provide new and fuller utilization of basement spaces a high degree of technical skill is required in their design and construction. A deeper understanding of moisture penetration and methods of compensating against its intrusion, improved ventilation and heat insulation are necessary if basements are to be used to their full capacity. However, waterproofing of basements is a complex problem and permanent watertightness is difficult to achieve.

This book, the third in a series on structural failures in residential buildings, fills a void in the practical implementation of basements as useful living accommodation and follows the pattern set in earlier volumes, of isolating a problem and providing a constructional solution to it. A large proportion of this book deals with drainage methods to reduce the ground water pressure adjoining basement walls and floors and will provide the architect and related professions with an invaluable manual.

John G. Roberts
WSA   UWIST
Cardiff, 1979

Contents

Preface

This report describes those areas and functions of underground building components which frequently fail and thus lead to structural damage, and derives recommendations for the prevention of these failures. As in the reports already published on other external components of buildings – Flat roofs, roof terraces, balconies and External walls and openings (Volumes 1 and 2 in this series) – the problems and causes of structural damage dealt with here were largely determined from a detailed enquiry into structural damage during the research project 'Problems of structural failure – prevention of structural damage in residential buildings', commissioned by the Ministry of the Interior of North Rhine-Westphalia.

The results of the investigation of damage to basements, adjoining land drains and foundations, which would have been impossible without the open cooperation of the publicly appointed building experts from the Chambers of Commerce, Industry and Craft in North Rhine-Westphalia, have been listed and published separately. Some of the facts which were discovered will, however, be briefly mentioned here, since they largely determined the specific aim and arrangement of this report.

Because 45% of the damaged basements were partially or wholly intended for living purposes or similar uses, particular value is placed on representing the problems and design possibilities of constructional forms which would satisfy more exacting damp-proofing and thermal protection requirements. In doing this, it is not the intention to encourage the constant use as living areas of basement rooms which would be unfavourable in physiological terms and inadmissible to the building inspection authorities. Instead, suitable methods are described for the situation which is found frequently where parts of the external walls of living rooms are underground (46% of the damaged buildings were constructed on a slope) and for the increasing number of basement rooms used on a non-habitable basis (e.g. hobby rooms, bars), or for the storage of moisture-sensitive goods; this has to be done because, often, building practice clearly no longer matches the changed requirements.

A further important result of the enquiry into structural failure was the discovery that by far the major part of the damage was caused by exposure to accumulated water – a type of defect which is particularly common in buildings constructed on slopes or on non-cohesive types of soil (as was the case with 78.8% of the damaged buildings). Suitable methods of protection against this exposure to water, which includes drainage measures, therefore occupy a particularly large proportion of this report. The lack of specialist literature to explain the methods for carrying out these protective measures stresses the expediency of treating this as a priority. Finally, inadequate investigation of the land water conditions that exist in the soil and the inadequate coordination between the tanking system and this water pressure represent the most frequently recurring design errors which lead to damage; these facts were decisive in planning the structure of the report.

Each of the main chapters, which are classified according to the structural components, is preceded by a separate section outlining basic considerations on the determination of the water level and pressure, and on the selection of the type of sealing employed. This results in a main classification as follows:

O – Determination of the water level and pressure, and of the sealing system
A – Adjoining land drainage
B – External basement wall
C – Basement floor

The priorities set from the above enquiry mean that no attempt has been made to describe all the possible designs for basement components in dealing with typical cross sectional designs and with the points of detail in separate problem areas. This publication therefore does not represent a complete guide on the construction of basement components but rather aims at preventing the damage which frequently occurs.

The treatment of the singular structural defects allocated to the individual problem areas was purposely not restricted to notes on how to remedy them. Instead, the recommendations for preventing structural defects are preceded by the respective damage descriptions, which justify the need for the measures suggested. There then follows a description of the relationships to soil mechanics, architectural physics, and materials and building techniques that emerged from an analysis of the damage, and required consideration. Only by providing an insight into the need for formal recommendations will it be possible to change procedures employed in the past. To make the information readily obtainable, the main points are summarised in lists. Bibliographical notes are also provided, for the reader who is interested in a deeper study of a particular problem area.

In describing the damp-proof measures used, terminology which was as standard as possible and which corresponded to current linguistic usage was followed. The coining of new words was avoided so as not to impede the reader's progress through the text. Damp-proof measures are therefore generally referred to as sealing measures, irrespective of their degree of efficiency and accordingly the terms 'sealing' and 'sealing layer' are used. Impervious plaster and concrete which fulfils a sealing role is thus referred to as 'water-proof plaster' or 'water-proof concrete' according to general linguistic usage.

The illustrations that accompany each defect show examples of damage and possible constructional solutions. The solutions are not intended as the correct alternative for the example of damage in each case; instead, the negative and positive examples should illustrate the contents of the text as typically as possible. With this in mind, the illustrations purposely avoid providing standard details, but give one of several possible solutions. In the majority of case studies there is no intention of discussing their fundamental impracticability in examples of damage, nor is it the aim to give the impression that a perfect solution is only possible with the materials and product shown in the illustrations. However, on the understanding of the conditions of each case and the recommendations derived from them, it is quite possible to check that materials, structural components and methods of construction offered are suitable for application in the example. The constructional recommendations are derived from the damage which was determined from an extensive evaluation of literature and product information,

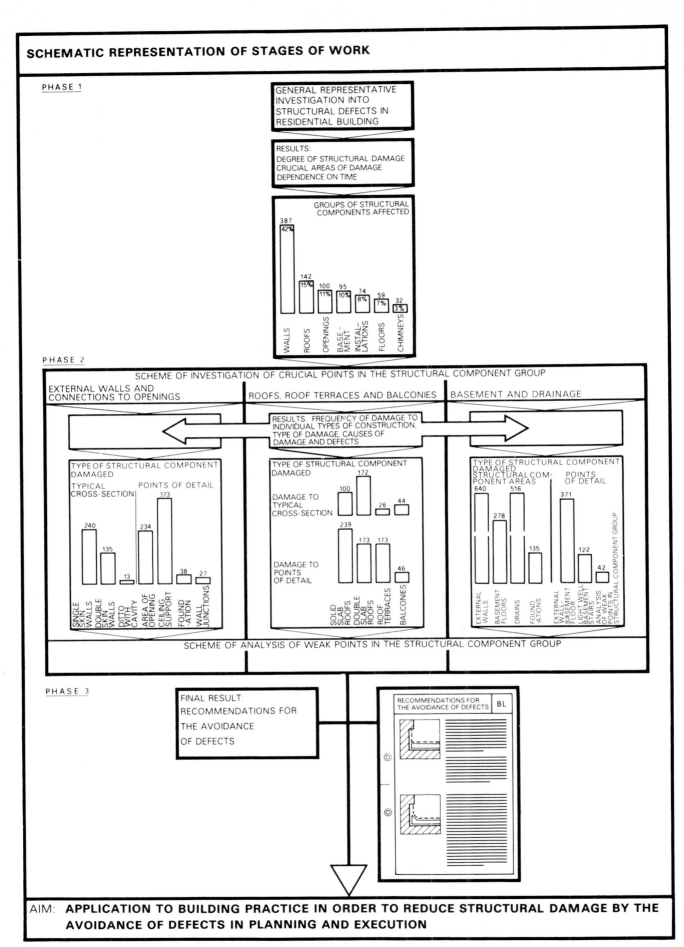

SCHEMATIC REPRESENTATION OF STAGES OF WORK

PHASE 1

GENERAL REPRESENTATIVE INVESTIGATION INTO STRUCTURAL DEFECTS IN RESIDENTIAL BUILDING

RESULTS:
DEGREE OF STRUCTURAL DAMAGE
CRUCIAL AREAS OF DAMAGE
DEPENDENCE ON TIME

GROUPS OF STRUCTURAL COMPONENTS AFFECTED

387 42% WALLS
142 15% ROOFS
100 11% OPENINGS
95 10% BASE-MENT
74 8% INSTAL-LATIONS
59 7% FLOORS
32 3% CHIMNEYS

PHASE 2

SCHEME OF INVESTIGATION OF CRUCIAL POINTS IN THE STRUCTURAL COMPONENT GROUP

EXTERNAL WALLS AND CONNECTIONS TO OPENINGS

ROOFS, ROOF TERRACES AND BALCONIES

BASEMENT AND DRAINAGE

RESULTS: FREQUENCY OF DAMAGE TO INDIVIDUAL TYPES OF CONSTRUCTION, TYPE OF DAMAGE, CAUSES OF DAMAGE AND DEFECTS

TYPE OF STRUCTURAL COMPONENT DAMAGED

TYPICAL CROSS-SECTION   POINTS OF DETAIL

373
240
234
135
38
27
13

SINGLE SKIN WALLS
DOUBLE SKIN WALLS
DITTO WITH CAVITY
AREA OF OPENING
CEILING SUPPORT
FOUND-ATION
WALL JUNCTIONS

TYPE OF STRUCTURAL COMPONENT DAMAGED

DAMAGE TO TYPICAL CROSS-SECTION

172
100
26
44

DAMAGE TO POINTS OF DETAIL

239
173
173
46

SOLID SLAB ROOFS
DOUBLE SLAB ROOFS
ROOF TERRACES
BALCONIES

TYPE OF STRUCTURAL COMPONENT DAMAGED
STRUCTURAL COM-PONENT AREAS   POINTS OF DETAIL

640
516
371
278
135
122
42

EXTERNAL WALLS
BASEMENT FLOORS
DRAINS
FOUND-ATIONS
EXTERNAL WALL-BASEMENT FLOOR
LIGHT WELL BASEMENT STAIRS
ANALYSIS OF WEAK POINTS IN STRUCTURAL COMPONENT GROUP

SCHEME OF ANALYSIS OF WEAK POINTS IN THE STRUCTURAL COMPONENT GROUP

PHASE 3

FINAL RESULT:
RECOMMENDATIONS FOR THE AVOIDANCE OF DEFECTS

RECOMMENDATIONS FOR THE AVOIDANCE OF DEFECTS   BL

AIM: **APPLICATION TO BUILDING PRACTICE IN ORDER TO REDUCE STRUCTURAL DAMAGE BY THE AVOIDANCE OF DEFECTS IN PLANNING AND EXECUTION** and from the author's experience in the research project. They reflect the present state of the construction industry. Despite this broad initial basis and careful preparation, the formulation of positive design and construction suggestions forces one to make decisions. The main criterion in making decisions was that of achieving damage-free constructions over prolonged periods even under unfavourable construction and water pressure conditions. In particular, recommendations for the damage-free design and execution of structural components in the ground must endeavour to achieve a high degree of safety. The cost of structural elements below ground during construction of the building is slight in comparison to the sometimes unusually high cost of subsequent repairs. In many examples the repairs remain inadequate, despite their high cost.

The present report aims to present the results of structural damage research in a form suitable for direct practical application. Not until the results of what occurs in practice have been understood and implemented can it be said that the present work has achieved its aim of preventing structural failure. The wide distribution and the great interest shown in the recommendations published in the last few years on the prevention of structural failure in flat roofs, roof terraces and balconies and in external walls and openings seem to confirm the effectiveness of this means of presenting the information. However, an active exchange of experience in the application of the results is desirable. We would therefore welcome critical appraisal.

Key to symbols used in the illustrations

Note: The illustrations represent principles and are not always to scale. In all illustrations, the outside of the structure is always represented by the left-hand (cross sections) or the lower side (outlines) of the picture.

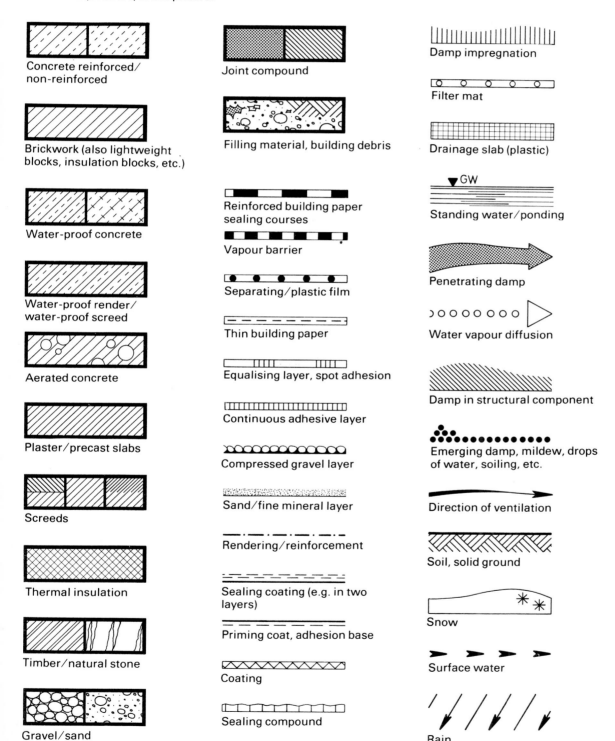

Concrete reinforced/
non-reinforced

Brickwork (also lightweight
blocks, insulation blocks, etc.)

Water-proof concrete

Water-proof render/
water-proof screed

Aerated concrete

Plaster/precast slabs

Screeds

Thermal insulation

Timber/natural stone

Gravel/sand

Joint compound

Filling material, building debris

Reinforced building paper
sealing courses

Vapour barrier

Separating/plastic film

Thin building paper

Equalising layer, spot adhesion

Continuous adhesive layer

Compressed gravel layer

Sand/fine mineral layer

Rendering/reinforcement

Sealing coating (e.g. in two
layers)

Priming coat, adhesion base

Coating

Sealing compound

Damp impregnation

Filter mat

Drainage slab (plastic)

Standing water/ponding

Penetrating damp

Water vapour diffusion

Damp in structural component

Emerging damp, mildew, drops
of water, soiling, etc.

Direction of ventilation

Soil, solid ground

Snow

Surface water

Rain

λ

Determination of the water pressure and level and of the sealing system

From building research surveys it can be concluded that under-estimating the actual water pressure on basement walls and floors and the resulting inadequate protection measures in the basement region are the commonest cause of extensive damp damage in basements.

Only in relatively few cases is the problem one of inadequate consideration of the groundwater level since this form of exposure to water under pressure, if not previously known, generally becomes clearly identifiable after the trial hole has been excavated: therefore an appropriate sealing system can be selected at this point if not sooner.

More often damage is due to defective protection methods being taken against exposure to water due to seepage which, especially in the case of cohesive soils and inclined sites, overstresses the external structures of the basement as water builds up. If the water is not removed by effective drainage measures and the basement structures have not been given any means of sealing in addition to protection against soil dampness failure occurs.

Damage in the case of pressure being imposed on the basement structures by soil moisture is observed in relatively few cases. The damage results mainly from a failure to match the sealing system to the use to which the basement is put, viz. very simple sealing measures are used for rooms with high requirements for the dryness of the surfaces of the structures.

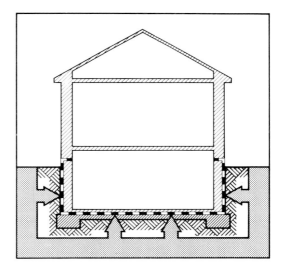

Early determination of the actual soil and water conditions and the stresses these impose on the basement structures is thus of fundamental importance in preventing structural damage since this is the only way in which a decision on an appropriate and economic sealing method can be reached. Especially in view of the expense of remedial measures, a detailed constructional method with a high degree of safety is of particular relevance for the basement region.

Consideration of the aspects outlined below relating to the determination of the type of stress and to specifying the possible sealing measures appropriate to the situation is therefore a basic prerequisite for subsequent design steps aimed at a trouble-free construction of the parts of the building below ground level.

Determination of water pressure and levels,
and of the sealing system

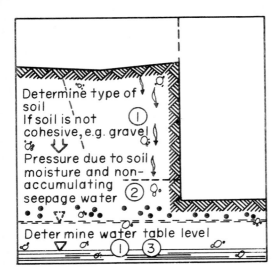

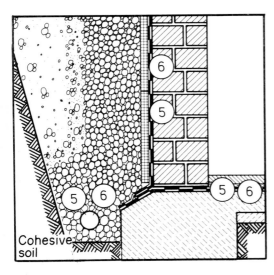

1 The sealing system has to be matched to the type of stresses and the use to which the basement is to be put. What water pressures are to be expected on the basement structures should therefore be elucidated at a very early stage in design (see 0 1.1.2).

2 Soil moisture can as a rule only be assumed in the case of very permeable, non-cohesive soils (gravel/sand), provided no groundwater is present. In examples of cohesive soils and inclined sites, allowance must inevitably be made at least for short-term build-up of seepage water (see 0 1.1.2).

3 If the form of pressure is not known from detailed experience on the building site of neighbouring building projects, information should be obtained, if necessary, through a soil expert on the level of the groundwater and on the type of soil and sequence of layers characterising the water conditions. Information on the groundwater level may be obtained, among others, from the National Water Board (see 0 1.1.2).

4 For pressure caused by **soil moisture** and **seepage water which does not build up,** the following sealing measures are suitable:
On vertical surfaces, e.g. basement walls:

- Three layer hot bituminous coatings or bituminous emulsion coatings,
- two layers of filler material,
- water-proof plasters,
- sealing compounds (see 0 1.1.3 and B 1.1).

At horizontal surfaces, e.g. basement floor:

- water-repellent screeds,
- sealing compounds with protective screed,
- single-layer water-proof felt with protective screed (see 0 1.1.3 and C 1.1).

5 In examples of **short-term build-up of seepage water** the sealing measures suitable for soil moisture can only be used if effective drainage is provided (see A 1.1). Under any other conditions the following measures must be taken to deal with this type of load on the wall and floor:

- surrounding tanking with water-proof concrete,
- water-proof felt in two layers with a protective screed (see 0 1.1.3, B 1.1 and C 1.1).

6 In examples of the **long-term build-up of seepage water,** sealing measures consisting of water-proof felt in two layers can be provided for the wall and floor if the water pressure is reduced by effective drainage (see A 1.1), otherwise sealing systems suitable for groundwater should be selected.

7 In examples of **exposure to groundwater** the following should be provided:

- water-proof felt in at least three layers,
- enclosing tanking with water-proof concrete manufactured by specialist building firms. The water-proofing of the basement in the case of pressures due to groundwater should not as a rule depend on drainage measures (see 0 1.1.3, B 1.1 and C 1.1).

Determination of the water pressure and level,
and of the sealing system

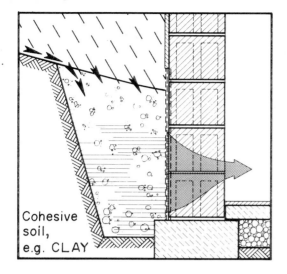

Cohesive soil, e.g. CLAY

The vast majority of examples of moisture penetration and water damage to basement walls and basement floors, which took the form of discolourations, formation of fungi, damage to paintwork and internal plasters, peeling of finishes, or even of leachings, run marks, and puddle formation, were observed because of soil conditions and inadequate sealing measures.

The buildings were erected in cohesive soils which were either impermeable or of low permeability, such as loam, clay or marl. Frequently considerable quantities of surface or seepage water were directed to the external basement structures as a result of the site being on a slope or of a large paved area drained in the direction of the building.

Effective drainage systems and seepage layers in front of the basement walls or beneath the basement floors were not constructed. The protective measures at the basement walls (e.g. two layers of bituminous asphalt) and at the basement floors (e.g. hard core) frequently did not even conform to a professional seal against moisture in the soil.

In a few examples, moisture penetration damage occurred as a result of moisture in non-cohesive soils.

Equally small was the number of examples of serious flooding in basements in which the water table rose above the top surface of the basement floor after prolonged heavy periods of rain when the basement floor and external walls did not have a water-pressure resistant seal.

The costs incurred in eliminating the damage described above were considerable and were many times the cost of initially introducing appropriate sealing and protective measures.

Points for consideration

– The construction of an appropriate sealing system requires an accurate knowledge of the pressures imposed by the water present in the soil, particularly in the back-filled areas directly adjacent to the basement wall and basement floor following initial excavation work.

– The basement structures can only be considered immune from the pressure of soil moisture (ground moisture) in the form of pressure-free, non-seeping, and non-dripping water which adheres to the soil particles (retained water) or rises up between them (suction water, capillary water) in examples of permeable, non-cohesive soils (gravel, sand). Equally, it is only in the case of these types of soil and of an appropriate permeable back-filling of the site that it can be assumed that seepage water originating from the surface of the land or from water-bearing strata will not build up in front of the basement wall (non-accumulating seepage water).

– In examples of the more common cohesive soils which are either impermeable or of low permeability (loam, clay, marl, silt) it must be assumed that surface and stratum water will seep into the back-fill material used in the construction working area (seepage water) and may build up for shorter or longer periods in front of the basement wall (water build-up). Once the type of soil is known (non cohesive or cohesive) – if the permeability coefficient is below 0.1 mm/sec the soil is cohesive – the probability of water accumulating or not accumulating can therefore be assessed.

– Whether a short-term, low-level or prolonged and heavy exposure to accumulated water is to be expected, depends on the permeability of the soil – i.e. on the routes by which the water can drain away – on the one hand, and on the other, on the influx of water. In examples of soils which are virtually impermeable (e.g. clay), or of large surfaces and strata draining towards the building (e.g. sloping site) long-term build-up of seepage water has therefore to be allowed for.

Determination of the water pressure and level,
and of the sealing system

– Groundwater above the base of the basement floor imposes water pressure on the structural parts of the basement wall. Because of the extensive design consequences of exposure to water under pressure, an exact knowledge of the groundwater table (particularly of the maximum level) is important. Experience on neighbouring building sites or data on groundwater table averaged over rather large areas can only be applied with sufficient safety to a particular building site if consideration does not have to be taken of variations in the height of the water table as a result of a fissured ground structure or irregular soil composition.

– Conclusions may also be drawn on the type of exposure through observations on the excavated trial hole. Thus, standing water due to precipitation indicates low soil permeability and consequently the possibility of water accumulation problems. If there is standing water even without rainfall in the trial hole, groundwater has to be reckoned with. The assessment of the problems to be expected on the basis of such observations is however uncertain, since weather conditions and variations in the water table can lead to false conclusions. An early, exact knowledge of the type of exposure also permits a better evaluation of the structural sealing measures needed and a more realistic estimate of the cost, and this may result in a lower expenditure on design.

– The costs of having an assessment and determination of the problems to be anticipated made by an expert, particularly if there is a possibility of constant accumulation of water or of groundwater, are small in comparison to the expenses incurred in eliminating damage in basements at a later date.

Recommendations for the avoidance of defects

● Even at an early stage in the design it is essential to elucidate whether water pressure is likely to be imposed on the structural parts of the basement by

– soil moisture and non-accumulating seepage water,
– short-term accumulation of seepage water in small amounts,
– prolonged accumulation of seepage water in rather large amounts,
– groundwater present at all or variable times.

● As a rule, soil moisture can only be assumed in examples of highly permeable soils (gravel, sand), provided no groundwater is present.

● In examples of cohesive soils and sloping sites allowance must inevitably be made at least for short-term accumulation of seepage water.

● If the exposure conditions are not known from detailed experience on the building site from neighbouring building projects, information should be obtained on the groundwater table and its variation, the permeability and the other data on the type of soil and layer structure that characterise the water conditions in the soil. Information on the level of groundwater and its variation is given, among others, by Government National Water Authorities.

When in doubt, a soil expert should be involved who will, if necessary, obtain more exact information from excavation or boring.

● Observations on the excavated trial hole should only be used as a supplement to the picture already obtained and are insufficient in themselves, for example for determining the possible groundwater table or the existence of stratum water.

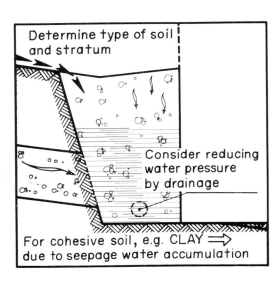

Determine type of soil and stratum

Consider reducing water pressure by drainage

For cohesive soil, e.g. CLAY ⟹ due to seepage water accumulation

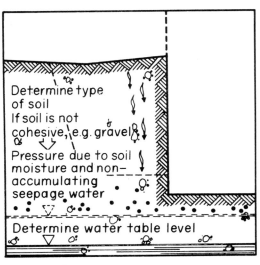

Determine type of soil
If soil is not cohesive, e.g. gravel

Pressure due to soil moisture and non-accumulating seepage water

Determine water table level

Determination of the water pressure and level,
and of the sealing system

Damp penetration damage to basement walls has frequently occurred when the following moisture protection measures have been carried out:

- one to three coats of bituminous asphalt on rendered or unrendered brick and concrete walls,
- water-repellent finishes on brickwork,
- sealing compounds with or without a first coat of rendering on brickwork and concrete.

Damp penetration damage mainly occurred in the basement floor if no protective or sealing layer was present in addition to a layer of gravel, ash or slag inhibiting capillary action below the floor slab.

The basement walls and floors with the protective measures described above showed damp damage because in the majority of examples accumulated seepage water imposed a temporary or permanent pressure on the structural parts, while in a smaller number of examples the problem was due to groundwater. A reduction in the pressure by drainage had generally not been undertaken or had been inadequately carried out.

Points for consideration

- The measures available for protecting structures in the ground against moisture offer resistance to the pressure of water in various degrees. On the one hand, this is due to the varying degree of imperviousness of a continuous surface of the sealing system, and on the other to the susceptibility to cracking and the chances of defects and damage arising from the construction.

- The smallest possible risk of damage is of considerable importance in examples of protective measures below ground level since to correct them at a later date is expensive or may not even be possible. Protective measures that provide low imperviousness in the surface and/or rather high susceptibility to cracking and to defects are therefore only suitable for areas of low water pressure. The sealing system must therefore be matched to the type of pressure.

- Multiple hot bitumen layers or bituminous asphalt coatings seal the floors of the basement only at the surface, while water-repellent renderings, screeds and sealing compounds to the walls are layers which have extremely few, narrow and capillary-inactive pores. However, for the reasons described above, these sealing measures, if carried out with reasonable care, only slow down the moisture penetration through the structure to an extent such that when the water pressure is low, the amount of water which can diffuse to the interior and evaporate is larger than the quantity coming in and adequate dryness is achieved for the structural parts of the basement if it is used conventionally. If the water does not penetrate into the structure by capillary action, but is forced through pores, fissures and cracks by an external standing volume of water (accumulated water, groundwater), rather large quantities of water may come through to the interior surface. Coatings, water-repellent finishes and sealing compounds are therefore only suitable for withstanding pressure from soil moisture and non-accumulating seepage water, and in examples of basements which are not used for permanent human occupation.

- One or two layers of sheet sealing materials such as heavy grade visqueen (polythene) or foils applied to the surface of the wall or floor are fairly water-tight as a continuous surface, but because of the possibility of defects in application or as a result of damage they are only suitable for temporary accumulations of water and for low pressure levels.

Determination of the water pressure and level, and of the sealing system

– Only multiple layers of sealing materials, which have been given a water pressure resistant backing, and tanking constructions consisting of water-repellent concrete which has been cast observing the requirements formulated for the manufacture of water-proof concrete, offer adequate certainty of protection against prolonged or constant accumulation of seepage water and groundwater.

– The water pressure on the structural parts of the basement resulting from accumulating seepage water and from groundwater can in principle be reduced by land drainage. If, as a result of the latter, sealing measures are used which are suitable for less intense water pressure, the protection afforded to the structural parts of the basement is only maintained if the operation of the drainage is trouble free. Efficient drainage, however, presupposes that a large number of factors are taken into consideration, requires a correspondingly high degree of care in design and execution, and is therefore associated with an increased risk of damage.

– In examples of heavy and prolonged exposure to water pressure, such as groundwater and large quantities of accumulating seepage water, land drainage should be rejected both because of the increased risk of its failure and because of the frequently resulting necessity of pumping out increasingly large quantities of water from sumps. The building authorities responsible may in any event refuse the installation of drainage systems to prevent a lowering of the groundwater level or an overloading of the main drainage system.

– In particular, in building situations where accumulation of seepage water is low and irregular, adjoining land drainage is a reasonable protection method which permits the application of sealing measures against soil moisture penetration through the structural parts of the basement.

Recommendations for the avoidance of defects

● The sealing system should be matched to the type of water pressure occurring and to the use for which the basement is intended.

● For water pressure resulting from **soil moisture** and **non-accumulating seepage water** the following sealing measures are suitable:

At vertical surfaces, e.g. basement walls:
– multiple layers of hot bituminous coating or bituminous asphalt coating, two layers of filler material, water-repellent rendering and sealing compounds.

At horizontal surfaces, e.g. basement floors:
– water-repellent screeds, sealing compounds with protective screed and a single layer of water-proof felt or heavy grade polythene with protective screed.

● In examples of **short-term accumulations of seepage water,** sealing measures suitable for soil moisture can only be used if effective drainage is provided. In any other circumstances, the following measures should be used for this type of exposure at walls and floors: water-repellent concrete tank structures or two layers of water-proof felt or heavy grade polythene with protective screed.

● In examples of **prolonged seepage water accumulation,** sealing measures consisting of two layers of water-proof felt or heavy grade polythene may be provided for the walls and floor if the water pressure is reduced by an effective drainage system, otherwise sealing systems suitable for groundwater should be chosen.

● In examples of **exposure to groundwater,** at least three layers of water-proof sheet or foil material and a tank structure made of water repellent concrete manufactured by specialist firms should be provided. The water-proofness of the basement should as a general rule not depend on drainage measures in examples of exposure to groundwater.

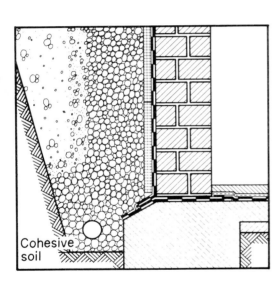

Cohesive soil

Specialist literature and guidelines

Grassnick, Arno; Holzapfel, Walter: Der schadenfreie Hochbau – Grundlagen zur Vermeidung von Bauschäden. Verlagsgesellschaft Rudolf Müller, Köln 1976.

Lufsky, Karl: Bauwerksabdichtung – Bitumen und Kunststoffe in der Abdichtungstechnik. 3. Auflage, B. G. Teubner, Stuttgart 1975.

Muth, Wilfried: Dränung zum Schutz von Bauteilen im Erdreich. In: Forum Fortbildung Bau, Forum-Verlag, Stuttgart 1977, Heft 8, Seite 115–127.

Reichert, Hubert: Sperrschicht und Dichtschicht im Hochbau. Verlagsgesellschaft Rudolf Müller, Köln 1974.

Schultze, Edgar; Muhs, Heinz: Bodenuntersuchungen für Ingenieurbauten. 2. Auflage, Springer-Verlag, Berlin 1967.

Terzaghi, Karl; Peck, Ralph B.: Die Bodenmechanik in der Baupraxis. Springer-Verlag, Berlin 1961.

DIN 4021, Blatt 1: Baugrund – Erkundung durch Schürfe und Bohrungen sowie Entnahme der Proben. Juli 1971.

DIN 4022, Blatt 1: Baugrund und Grundwasser – Benennen und Beschreiben von Bodenarten und Fels. November 1969.

DIN 4023: Baugrund- und Wasserbohrungen – Zeichnerische Darstellung der Ergebnisse. September 1975.

DIN 4031: Wasserdruckhaltende Bituminöse Abdichtungen für Bauwerke – Richtlinien für Bemessung und Ausführung. November 1959.

DIN 4095: Baugrund – Dränung des Untergrundes zum Schutz von baulichen Anlagen. Dezember 1973.

DIN 4117: Abdichtung von Bauwerken gegen Bodenfeuchtigkeit – Richtlinien für die Ausführung. November 1960.

DIN 18 195 – Teil 4 (Entwurf): Bauwerksabdichtungen, Abdichtung gegen Bodenfeuchtigkeit, Ausführung und Bemessung. November 1977.

DIN 18336 – VOB, Teil C: Abdichtung gegen drückendes Wasser. Oktober 1965.

DIN 18337 – VOB, Teil C: Abdichtung gegen nichtdrückendes Wasser. Februar 1961.

Problem: Adjoining land drainage

Land drainage systems are intended generally to prevent any form of water accumulation in front of building structures. They are necessary when the structures in the ground are not themselves protected against exposure to accumulated water.

The land drain – the most important functional element in the drainage system – is a pipe which as a rule runs round the foundation area of the whole building in the form of an uninterrupted ring; it is particularly intended to receive water accumulating and building up in front of parts of the building and in some examples to absorb water fed to it from vertical seepage layers or from surface drainage systems below the floor slab, and at an even gradient to carry this surface and land water into the main drainage system. This significant role is in contrast to the view commonly held that any pipe laid 'somehow or other' in the ground is already an efficient drain.

In principle, a drainage system can only be dispensed with if the building is either erected in rapidly draining, uniform soils in which no horizontal movement of the precipitation or surface water percolating through the soil is to be expected or if a form of moisture exclusion has been selected which is matched to the type of water pressure. In all other examples, particularly if there are no clear-cut circumstances whose effect can be assessed relating to the various determining factors (rainfall, topography, soil conditions), a permanently effective drainage system should be laid with an even gradient which has been planned at an early stage in the design since it may in some instances affect the load-bearing characteristics (if it runs under the foundation level) of the building structure above.

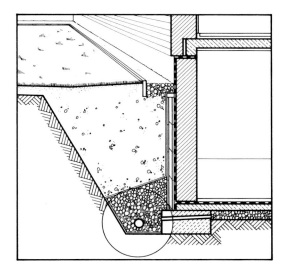

In over half the buildings included in the survey, drainage systems had been laid which were entirely or partially inoperative, and had consequently resulted in serious failures. In addition, there were a large number of examples of failure which had been caused by the complete lack of a drainage system. In this connection, the central issue was that about 91% of all the drainage failures were due to the fact that the pipes of the land drainage system were either ineffective or of limited efficiency due to inappropriate position, gradient, dimensions, etc.

The causes of failure deduced from the examples of damage and the recommendations evolved for dealing with them indicate the importance of drainage pipes as part of the entire drainage system and show that detailed design and execution is needed in this area.

Drainage

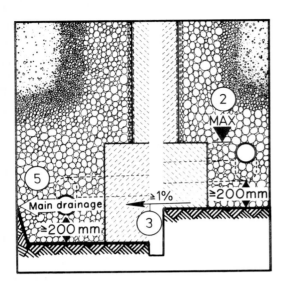

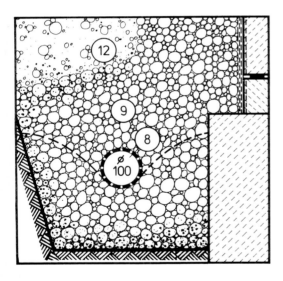

1 The land drainage must surround all the parts of the building subject to exposure including all projecting and receding parts and must run with as uniform a gradient as possible without interruption from the uppermost point to the lowest point (sump). At all changes of direction, inspection pits should be installed which permit cleaning (flushing out) of the drainage pipe (see A 1.1.2).

2 The drainage pipe should not be above the top of the foundation at its highest point. The position of the drainage pipe at the lowest point is given by the location of the pipe at its maximum level (top of foundation) less the gradient depth, taking account of the pipe thickness needed. If necessary, the foundations should be stepped so that there is no risk of their base being undermined. The position of the adjoining land drainage is also dependent on whether drainage is installed below the basement level (surface drainage). The latter has to be fed by suitable connecting pipes laid to a gradient through the foundation to the continuous land drain system (see A 1.1.2 and A 2.1.6).

3 For situations arising in most buildings and if the drainage is laid with the greatest care, a gradient of 0.5% is quite adequate, but a gradient of at least 1% should be aimed at. In examples of large building projects which require long drainage pipes, of unusual water and soil conditions (e.g. occurrence of stratum water) and of all special examples where high discharge rates are to be expected [$\geq$ 0.1 l/sm (litre per sec per linear metre)], the layout, gradient and nominal diameter of the drain pipe should be specified by a drainage specialist (see A 1.1.2).

4 Drain pipes laid without a gradient are at best only suitable for small building projects with short pipe lengths and low water influx (approx. 0.05 l/sm). In all examples it is essential to ensure that the pipe remains free of any debris that might obstruct the discharge (see A 1.1.2).

5 In designing and specifying the gradient and position of the drainage system, it is necessary to start from the point of connection to the main drainage system. Because of the danger of back-surge the entire drainage system should be situated above the highest possible water level in the main drainage system (see A 1.1.2 and A 2.1.3).

6 Earthenware drain pipes should be so laid that they have a joint of 1-5 mm depending on the water influx and the type of percolation. If this cannot be maintained with certainty (if necessary, using insert rings or drain sleeves which also have a filtering action), slotted or perforated drainage pipes (e.g. made of stoneware, concrete or plastic) or pipes of porous concrete should be provided (see A 1.1.3 and A 2.1.2).

7 Use should only be made of pipes which have a smooth internal bore and which can take up the incident water on all sides (fully porous land drain pipes) (see A 1.1.3).

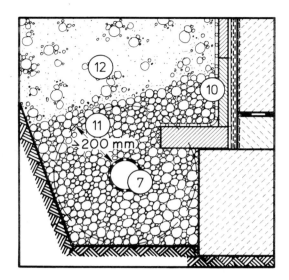

8 As a rule, pipes with a diameter of 100 mm can be considered adequate. For unfavourable soil conditions (e.g. surface or underground flow directed towards the building, groundwater, extremely large catchment area) and where water is fed in from the surface drainage, the cross-section should be increased (see A 1.1.3).

9 The drainage pipe should be encased over its full extent by a material with stable filtering properties. Suitable materials are, for example, gravel (concrete, aggregate material, broken brick or granulated slag) which either has itself the correct grain structure for filtering or is protected against the penetration of fine soil particles by a further filter layer (e.g. filter matting made of artificial fibre) (see A 1.1.4).

10. Filter mats, which must have stable filtering properties with respect to the surrounding subsoil or back-filled ground, should always be provided if there is any uncertainty about the nature (grain composition, size, distribution) of the surrounding subsoil or back-filled soil and of the material used for packing around the pipe. They should always be installed without flaws or gaps (if necessary, with overlapping of the joints) between the soil and the packing material (see A 1.1.4).

11 The packing and filtering material must surround the drain pipe on all sides to a thickness of at least 200 mm. If coarser materials are used and the thickness is less, filtration stability should be guaranteed by a filter fibre mat which surrounds the main packing on all sides (see A 1.1.4).

12 The provision of the filter packing material, as well as the back-filling of the site, should be undertaken as soon as possible after the drain pipes have been laid; all the joints, the gradient, and the position of the pipe should be checked beforehand to see that they are to specification (see A 1.1.4).

Drainage
Adjoining land drainage

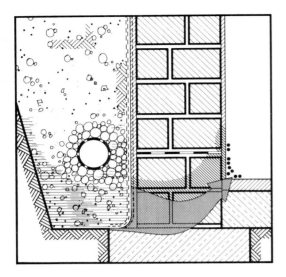

Insufficient gradient and incorrect positioning, i.e. errors in laying the surrounding land drain pipes, were frequent causes of failure or, at least, had an intensifying effect on damage.

The pipes were located at the level of the basement floor, some were well above the level of the basement floor at their highest point and some had been laid with no gradient or with a fall in the wrong direction. In examples where surface drainage had been provided below the level of the basement, it could not be connected in to the surrounding system. The consequent damage took the form of damp penetration of the external walls of the basement, especially at the connection to the foundations.

Points for consideration

- Over the whole drainage system, long-term accumulation of water may occur at the base of the site depending on the quantity of water seeping through and on the resistance to it entering the land drain system. This risk occurs especially in the region below the drainage pipes (between the underside of the pipe and the subsoil) if cavities are present or unsuitable back-filling materials which are not capable of seepage (e.g. cohesive soils, concrete and mortar residue) only permit the water accumulating in this region to enter the drain pipe to a limited extent. If this zone is level with walls or floor which cannot be exposed to water under pressure, the risk of damp penetration is especially great.

- The water disperses in the drain pipe as a result of the height of the water level. If the drain pipe is not inclined itself, only the difference in the level of fill is effective, whereas in examples where lengths of pipes have been laid with a gradient the two are cumulative. One may regard 0.5–2% as an effective drop which also allows for the building tolerances that may arise during erection. From this it follows that for a standard length of a drainage system between the highest point and the outlet to the main drainage (collecting sump) of about 20 m there should be a difference in height of 100–400 mm. This difference in levels which is needed for proper functioning may, on the one hand, lead to too great a height at the upper point (above the level of the basement floor), and on the other hand to undermining of the foundations at the lowest point. In designing the foundations, all the provisions for drainage must therefore be taken into consideration, e.g. possibly by increasing the depth of the foundations.

- Drain pipes without or with insufficient gradient increase the risk of damage due to inadequate drainage capacity.

- Level drain pipes installed above the projection of the foundation slab but adjoining walls liable to seepage are indeed capable of permitting the water to run away as a result of the drop in water level established, but the water may build up at the base of the outer basement wall as a result of the entry resistance in the pipes or the pressure necessary for penetration and this may, especially if there are flaws in the tanking, result in damp penetration. In addition, the pipes frequently have no gravel bedding so there is also a danger of silting up. This causes a worse problem since the soil particles can only be flushed out with difficulty.

Drainage
Adjoining land drainage

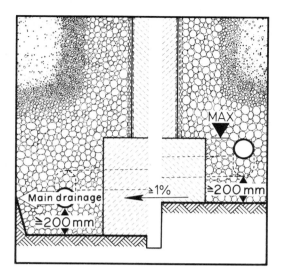

Recommendations for the avoidance of defects

● To be certain of preventing moisture penetrating into the outer basement wall as a result of temporary build-up of water over the length of the drainage system, the drain pipe and its stop end should not be above the top of the foundation (i.e. the wall-base surface or the top of the basement slab in general) at its highest point.

● The land drainage must surround all parts of the building under water pressure including all projections and receding parts and run with as uniform a gradient as possible and without interruption from the uppermost point to the lowest point (collecting sump). At all changes of direction, inspection pits should be provided which make it possible to clean (flush out) the drainage system.

● For the situations usually encountered in building and with careful laying of the drain, a slope of 0.5% is quite sufficient, or if allowance is made for building tolerances or to ensure a minimum gradient in the direction of the collecting sump or main drainage system, a gradient of about 1% should be aimed at. In examples of large building projects which necessitate long drainage systems, of unusual water or soil conditions (e.g. occurrence of stratum water) and of all special cases in which high discharge rates (>0.1 l/sm) are to be expected, the layout, gradient and nominal diameter of the drain pipe should be specified by a drainage specialist.

● Drain pipes laid without a gradient are at best only suitable for small building projects needing short pipe lengths and having a small influx of water (about 0.05 l/sm). In all examples it is essential to ensure that the pipe remains free of any debris that might obstruct the discharge.

● The position of the drainage pipe at its lowest point is given by the location of the pipe at its uppermost point less the depth of gradient, taking account of the necessary pipe thickness. If necessary, the depth of the foundations must be stepped so that there is no risk of their being undermined.

● In designing and specifying the gradient and position of the drainage system one should start from the point of connection to the main drainage system. Because of the danger of back-surge the entire drainage system should be situated above the highest possible water level in the main drainage system (see A 2.1.3).

● The position of the adjoining land drain is also dependent on whether drainage is installed below the level of the basement (surface drainage). The latter must be connected to the main drainage system by suitable connection pipes laid to a gradient through the foundation (see A 2.1.6).

Drainage
Adjoining land drainage

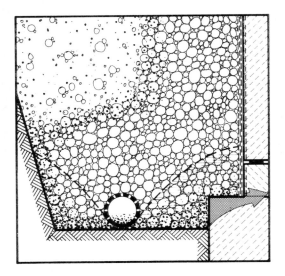

In some cases, the drainage pipes had a diameter (nominal width) of only 70–80 mm, or the pipes were only perforated on their upper side (fully porous land drain pipes) so that they had a correspondingly poor drainage capacity, associated to some extent with silting-up. When the moisture protection was defective, this led to damp penetrating the external basement walls over the length of the pipe system as a result of excessive water accumulation over the site.

Points for consideration

– The water percolating through the back-fill material surrounding the drainage pipe or flowing into that area and accumulating on the base of the original excavation enters the drainage pipe as a result of the drop in pressure. Along with the water, a certain amount of fine soil flows into the pipe; the amount depends on the diameter and number of the water entry apertures and the design and quality of the surrounding filtering material. These particles (especially if the gradient is inadequate) accumulate at the bottom of the pipe and reduce the effective bore of the pipe. In this connection, pipes made of earthenware are especially at risk at their joints (see A 2.1.2) while drainage pipes with slotted, perforated or porous casing surfaces (concrete and plastic) are exposed to the risk of silting up over their whole length.

– Rough or uneven finishes to the insides of pipes also reduce their drainage capacity.

– The flow into the drain pipe takes place from underneath. Pipes which are only perforated on their top side (land drain pipes porous on one side) or which are laid directly on the base of the original excavation, namely on solid soil, are not able to take up so much water, are therefore less efficient and more prone to silting up. They require a greater head of pressure to take up the water and the level of the water in the drainage trench remains higher than is the case for fully porous land drain pipes which can take up water from underneath.

– The nominal diameter of the pipe in cross-section to be selected is dependent on the influx of water and the amount to be drained away. The following factors influence the water influx: the situation of the terrain (e.g. incline), size of the catchment area, and structure of the soil; the latter may involve for example stratum water from remote areas, and the connection of surface drainage or ground water temporarily present.

Recommendations for the avoidance of defects

● Earthenware drainage pipes should be so laid that they have a gap of 1–5 mm at their joints depending on the water influx and the nature of the seepage or filter material. If this cannot be maintained with certainty (if necessary, using insertion rings or drain pipe sleeves which also have a filtering action), slotted or perforated drain pipes (made of stoneware, concrete or plastic) or pipes made of porous concrete should be provided.

● Only pipes which have as smooth an internal bore as possible and which can take up the incident water on all sides (fully porous land drain pipes) should be used.

● As a rule, pipes with a diameter of 100 mm can be considered adequate. For especially unfavourable terrain and soil conditions and if water is fed in from surface drainage, the cross-section should be increased if an exact calculation does not indicate that it is adequate.

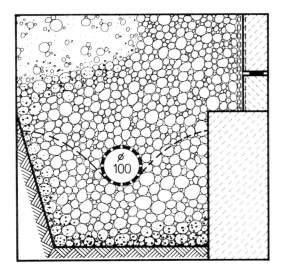

If the drain pipes had not been surrounded with a filtering and seepage layer (e.g. were laid directly on the base of the site exposed by the excavation or covered directly with back-filled soil), they generally exhibited drainage faults after a short time. This to some extent resulted in considerable damp penetration damage, particularly in the base area of the walls which were protected only against non-accumulating water. In some cases, damage arose despite the pipes being placed in a 'seepage layer' made of gravel, hard core or granulated slag. In this instance the grain structure or the thickness of the material was insufficient so that there was no effective filtering action. Most of the drainage systems consisting of earthenware and plastic pipes were affected.

Points for consideration

– The water flowing through the soil to the drain pipe carries soil particles which get into the pipe through the apertures needed for the influx of water (joint gap, perforations, slots) and this can lead to blockages due to increasing sanding or silting.

– For permanent efficiency of the drainage system, it is necessary for the water to get into the drainage pipe as quickly and with as little impedance as possible, i.e. the highest possible water permeability should be provided for the pipe and the surrounding material, but an adequate filtration stability of these layers is needed to prevent the drain pipe and the seepage material layer becoming silted up with water-borne soil particles. For all types of soil with a high content of fine particles, this danger should be prevented by special measures.

– Seepage layers consisting of grain sizes that are not graded or are only slightly graded are not suitable to retain fine soil particles in the long term since their filtering properties are not stable. The only materials that should be used are those with stable filtering properties capable of retaining fine particles of the soil to be filtered off (surrounding subsoil or back-filled material) in adequate amounts as a result of their pore structure.

– The filter function can also be taken over by single layers (e.g. filter mats, filter fibres) which must have sufficiently fine texture or pore structure to prevent the penetration of fine soil particles into the seepage material and the drain pipe while having adequate water permeability.

– The filter material should be uniformly graded and similar in form to the grain quality of the soil to be filtered.

– Water flows towards the drain pipe from all directions, i.e. including from below; this fact must be taken into account in laying the pipe by placing material with stable filtering property beneath the drain pipe adjacent to the underlying soil.

– Until the filtering and seepage material has been installed the pipes of the land drains are unprotected in the open excavation and they are therefore exposed to the risk of silting up (e.g. by loose soil particles carried into the pipe by rainfall).

Drainage
Adjoining land drainage

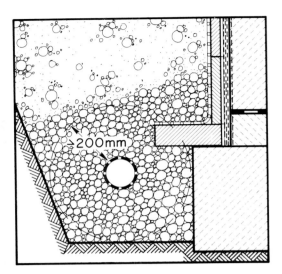

≥200mm

Recommendations for the avoidance of defects

● The drain pipe should be surrounded by a seepage material with stable filtering properties. Suitable materials are, for example, gravel (gravel for concrete with a leachable component content of diameter ≤ 0.02 mm ≤ 2–3 wt. %; diameter ≤ 0.063 mm ≤ 10 wt. %, or perhaps of broken brick or granulated slag) which has either a suitable filtering structure in its grain composition within itself or is protected by a further filter layer (e.g. filter mat of artificial fibres) against penetration by fine soil particles.

● Filter mats, which must have stable filtering properties with respect to the surrounding subsoil or back-filled material, should in any event be installed if there is uncertainty about the nature (grain composition, size, distribution) of the surrounding subsoil or back-filled material and about the seepage layer around the pipe. They must be laid without cracks or gaps (if necessary, with overlapping of the joints) between the soil and the seepage layer.

● In spite of a correct filtration structure, care has to be taken of the penetration of silt in the transition regions from the seepage layer to the surrounding subsoil or back-filled material. The seepage and filter material must surround the drain pipe to a thickness of at least 200 mm. If coarser materials are used and the thickness is less, the filter stability must be guaranteed by a filter fibre surrounding the seepage layer on all sides.

● The percolation and filtering layer, as well as the back-filling of the excavation, should be done as quickly as possible after the drain pipes have been laid to prevent silting up of the unprotected pipe by washed-out earth during rainfall. All joints, the gradient and dimension of the pipes should be checked beforehand to see that they are to specification.

Problem: Drainage below basement slab

If the water and soil conditions are such that water can penetrate the building from below, thereby giving conditions that might cause damage to the floor area of the basement, surface drains may be arranged beneath the basement floor to collect the water and feed it to the surrounding land drainage, thus saving the cost of more expensive sealing measures in the basement floor area. If, for example, the groundwater level is slightly below the basement floor, if there are changes between fast and slow draining (non-cohesive and cohesive) soils parallel to the basement walls which permit surface water to seep up to the floor of the basement or to collect above it, or if water-bearing soil strata are excavated in such a way that stratum water released in this way induces pressure on the basement slab from below, the capacity of the surface drainage will have to be designed for variable pressures by selecting appropriate constructional methods. This means a simple layer of gravel material with a permeable grain structure with stable filtering properties, or drainage with pipes.

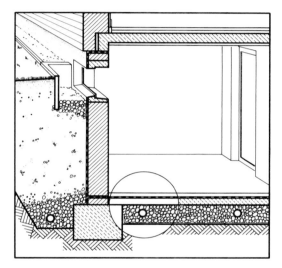

It is impossible to say with certainty whether damp penetration in the area of the basement floor can be attributed to incorrectly designed or constructed surface drains, since the independent factors which influence this are numerous (water pressure, type of soil, type of sealing, etc.) and since subsequent inspection and checking of the installed drains is unusually expensive and difficult.

Those surface drains which were found to be inadequate were either of drain pipes laid in gravel layers or of drains which did not use pipes. In addition, there were gravel layers or coarse gravel material which were referred to as surface drains even though they had neither stable filtering properties nor any apparent drainage action. Thus, the most they did was to inhibit the capillary flow of water.

The following descriptions and illustrations show possible ways of avoiding common faults in drainage below the basement slab and the damage which they cause to structural components of basements.

Drainage

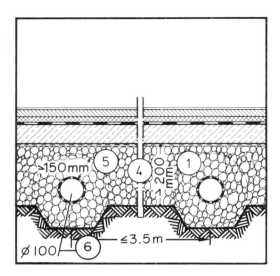

1 For drainage below the basement slab, a material should be chosen whose filtering properties are as stable as possible in relation to the surrounding soil. In most cases, graded gravel specified for concrete is suitable. The proportion of material that can be washed out of the gravel used should not, however, be too high (diameter $\leq$ 0.02 mm, $\leq$ 2–3 wt. %; diameter $\leq$ 0.0632 mm, $\leq$ 10 wt. %—0) (see A 1.2.2).

2 In the case of seepage layers made of 'small grain' gravel and of all types of fine-grain soil types, filter layers (e.g. fibre filter mats) should be arranged between drainage layers and the surface of the soil (see A 1.2.2).

3 Drains arranged below basement floor slabs must have permanently effective drainage to the outside (e.g. to the adjoining land drain) (see A 1.2.3 and A 2.1.6).

4 Drainage layers which do not use pipes must be made of a porous material with stable filtering properties at least 200 mm thick (see A 1.2.3).

5 If drains with pipes are to be used, the drainage pipes should be laid in such a way that at least a 200 mm thick permeable seepage layer remains between the pipe base and the surrounding soil. In order to ensure faster water absorption and discharge in the case of larger areas to be drained, it is advisable to lay the pipes in trenches with a suitable cross section and gradient so that the pipes are surrounded on all sides by a sufficiently thick seepage and filter layer (approx. 150–200 mm) (see A 1.2.3).

6 The drainage pipes themselves should be laid at maximum centres of 3.50 m and with a gradient of at least 0.5% to the collecting pipe and must be connected to this pipe at an acute angle to the direction of discharge. The nominal diameter of the pipes should be no less than 100 mm (see A 1.2.3).

Drainage
Drainage below basement slab

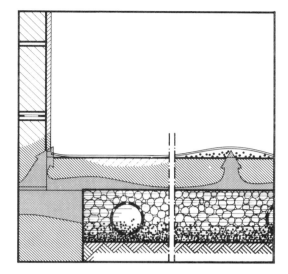

In many cases, drains laid below the basement floor for collecting water present under the building and for discharging it into the adjoining land drainage failed to operate correctly because of the incorrect choice of materials. Pipes which were laid partly in granulated slag material and partly in fine-grained grit layers frequently became blocked after a short time, so that the water remained beneath the basement floor and was able to penetrate into the building. Moreover, if there was no sealing layer above the floor slab, the water was able to enter inside the room freely and to damage floor coverings and furniture.

Points for consideration

– As in the case of the surrounding land drainage, with drainage below the slab the risk of clogging and thus the effectiveness of the drainage system is determined by the size of the openings through which water enters the pipes (joints in the case of earthenware pipes, size of the openings in the case of plastic pipes) and depends on the type and composition of the seepage or filter material surrounding the pipes (see also A 1.1.4).

– The drainage is more effective the coarser the pores of the back-fill layers or seepage layers whilst providing the same degree of stability against soil particles that are washed in (stable filtering properties). On the other hand, drainage layers made of fine grained seepage material allow the water to enter the drain pipe slowly and these become clogged with fine particles, so that the drains may be rendered useless after a short time.

– Although layers made of coarse gravel of uniform grain size allow the water to enter the drainage pipe quickly, they generally do not have stable filtering characteristics in relation to the surrounding soil so that, here too, fine particles can be carried into the pipe with the water and these can easily cause a blockage, particularly where the gradient is slight.

Recommendations for the avoidance of defects

● The material selected for surface drainage should possess filtering properties which are as stable as possible in relation to the surrounding soil. In most examples, graded gravel specified for concrete is suitable. The proportion of material which can be washed out of the gravel used, however, should not be too high (diameter ≤ 0.02 mm, $\leq$ 2–3 wt. %; diameter ≤ 0.063 mm ≤ 10 wt. %).

● In the case of seepage layers made of 'small-grain' gravel and where surrounding soil types are fine-grained, filter layers (e.g. fibre filter mats) should be arranged between the drainage layers and the soil surface.

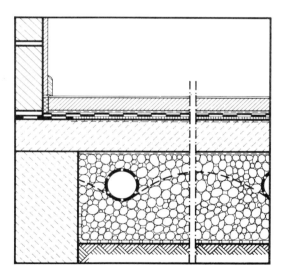

Drainage
Drainage below basement slab

Drains below basement slab which were not connected to the adjoining land drains, and whose slag or gravel layers were insufficiently thick and in which the pipes were laid directly on the surrounding soil, or which were laid without pipes despite heavy water pressure were not able to collect and discharge water flowing under the building in sufficient quantities, thus resulting in varying degrees of damp penetration through the basement floor slabs to the inside of rooms.

Points for consideration

– The water collecting beneath the base slab can lead to fairly severe damp penetration, particularly when it accumulates and thus exerts fairly prolonged hydrostatic pressure on the base slab.

– Drainage layers can only remain effective on a permanent basis if they are sufficiently thick, besides having a grain composition which is such that the water can filter and seep through. Because of the inevitable penetration of fine particles of soil, thin layers tend to clog completely in their cross section. With thicker layers a sufficient, effective layer for discharging the water remains, despite fine soil particles washed in through the transitional area. Moreover, after a time, a natural filter layer builds up between the surrounding soil and the seepage layer.

– In examples of heavier water pressures (e.g. groundwater which frequently rises for short periods), gravel and slag layers are inadequate forms of drainage on their own. Because of their irregular cavity distribution they have a relatively high flow resistance and are thus unable to discharge the water as quickly as pipes embedded in such layers can.

– Pipes laid on the excavated base, which are in contact with the surrounding soil, will be flooded with fine soil particles, since they are not provided with a protective filter layer, particularly on their undersides.

Recommendations for the avoidance of defects

● Surface drains laid below floor slabs must have permanently effective discharge to the outside (e.g. to the adjoining land drains) (for details see A 2.1.6).

● Drainage layers without pipes should be made of porous material, with stable filtering characteristics, at least 200 mm thick.

● If drains are to be fitted with pipes, the drain pipes should be laid in such a way that there is a permeable seepage layer at least 200 mm thick between the pipe base and the surrounding soil. In order to ensure faster water absorption and discharge in the case of larger areas to be drained, it is advisable to lay the pipes in trenches with a suitable cross section and gradient so that the pipes are surrounded on all sides by a sufficiently thick seepage and filter layer (approx. 150–200 mm).

● The drain pipes themselves should be laid at maximum centres of 3.50 m and with a gradient of at least 0.5% to the collecting pipe and must be connected to this pipe at an acute angle to the direction of discharge. The nominal diameter of the pipes should be no less than 100 mm.

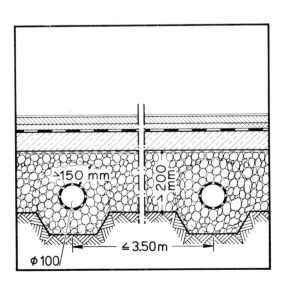

Problem: Vertical seepage and filter layer

The vertical seepage layer (drainage layer) in a section of the complete drainage system has the work of collecting water from the surface of the back-filled site and of quickly transporting it downwards to the drainage pipe. In order to ensure it remains permanently operative, particles of soil must be prevented from entering the seepage layer.

The vertical seepage layer should prevent the temporary or prolonged presence of water which can accumulate from percolating surface water and from stratum water, especially if, as a result of the size and position of the catchment area (on a slope) or of the nature of the soil strata, the quantity of water flowing towards the building is large and if this water accumulates in examples of cohesive surrounding soil or in examples of back-filling material which has cohesive components.

Besides a seepage function and a filter function where the surrounding or back-filled soil types are appropriate, the vertical drainage layer can also provide the wall sealing layer with effective protection against mechanical loads and damage during back-filling and against the harmful effects of sunlight during building.

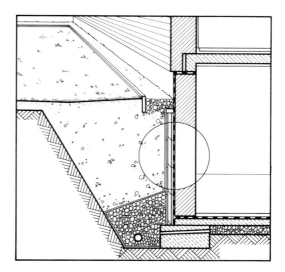

Despite their great importance in producing a functional drainage system, actual drainage walls which should fulfil the above-mentioned functions are rarely constructed and were equally rarely found to be harmful or to be the cause of damage. Layers of loosely packed seepage material made of concrete, and gravel layers in front of the external wall are the most common type of seepage layers, whilst drainage layers made of plastic sheets are very seldom used. Types of construction using loose large stone slab layers or corrugated sheets made of asbestos cement or bituminous materials – which are often called drainage layers – cannot be considered as seepage and drainage layers. At best, they are suitable as protective layers (see B 1.2.6).

The defects described here are thus confined mainly to constructions with seepage rubble and gravel layers as are recommendations for their design and construction.

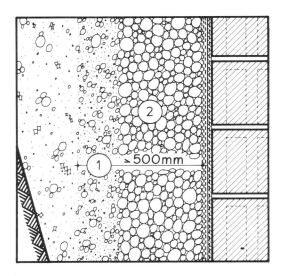

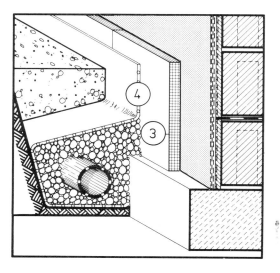

1 Vertical drainage layers, which must surround the whole of the walls under pressure without interruption, should be made sufficiently thick to ensure that an adequately large cross section remains clear for discharging water after taking into account the inevitable silting of the outer areas. If the whole of the excavated site is not back-filled with permeable gravel up to the drainage pipe, gravel layers with a minimum thickness of 500 mm should at least be built up in front of the external basement wall (see A 1.3.3).

2 If gravel layers are to remain permanently effective as drainage layers, their grain size should have suitably stable filtering properties in relation to the surrounding soil. To achieve this, the layer can be built up progressively to form a stage filter [e.g. from a 300–400 mm thick layer of coarse gravel (diameter 20–60 mm) and a 100–200 mm thick layer of sand (approx. diameter 0.2–4 mm), depending on the surrounding soil type] directly in front of the external wall or to form a mixed filter. For example, suitable materials for a mixed filter are graded gravel suitable for concrete or grit containing the smallest possible proportion of particles likely to be flushed out (see A 1.3.2 and A 1.3.3).

3 Drainage walls constructed directly in front of the external basement wall should take the form of dry-laid seepage materials (sheets) made of porous concrete or rubble or drainage sheets made of polystyrene or polyethylene foam. When using plastic drainage sheets, their compressibility and durability, which are important factors determining their permeability characteristics, should be taken into account by employing suitable measures (e.g. greater thicknesses as the height of the drainage wall increases) (see A 1.3.2 and A 1.3.3).

4 Filter layers (e.g. plastic or glass fibre mats) should in principle always be laid as a protection against silting if the walls to be drained themselves do not have sufficiently stable filtering characteristics (e.g. seepage rubble, gravel layers with grain structure which is only slightly graded or plastic sheets) or in the presence of very fine-grained soils. The filter layer must cover the whole of the basement wall to be drained without gaps (overlap $\geq$ 100 mm at the joints) and without defective points (see A 1.3.3).

Drainage
Vertical seepage and filter layer

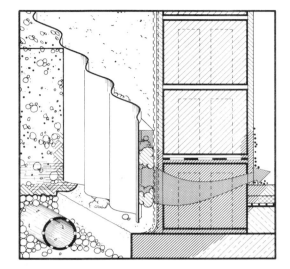

Even a short time after buildings were completed, loose rubble layers made of high porosity bricks or perforated sand-lime bricks, and asbestos cement corrugated sheets or bituminous materials, which were designed to feed the water to the drainage-pipe in front of the basement wall, were inoperative. The water to be discharged accumulated because of the silting of the rubble layers, of the cavities in back-fill material and particularly of the hollow spaces between the corrugated sheets and the basement wall, in many cases as the water reached up to the top or at least the centre of the wall. Especially after heavy rainfall and where walls were insufficiently sealed against the water pressure against them this resulted in signs of damp on the inside of the walls. Sometimes these were limited to isolated patches, but in most examples they were evenly distributed over large areas of the whole wall.

Points for consideration

– Depending on the type of surrounding soil and the back-filling material used (e.g. cohesive or non-cohesive soil, type of sealing, other materials which might impede water drainage), water was able to accumulate in front of the external wall for relatively long periods. The risk of the temporary accumulation of water must be minimised as far as possible, particularly in the case of walls which were only sealed against soil moisture. This can be achieved by means of porous layers in front of the external basement walls.

– Layers, which form hollow cavities either in cross section (e.g. loosely piled rubble walls) or directly in front of the external wall (e.g. corrugated sheets or other profiled protection sheets) may be unable to fulfil a permanent drainage function on their own. Because of their material construction they are unable to absorb water which collects through their surfaces (in some cases water can only pass through the joints and, if they are not protected, particles of soil can also be carried through) and to feed the water downwards to the drainage pipe. Because of the lack of a drainage construction, there is a risk of silting caused by fine soil particles and ultimately this will hold the water and allow it to accumulate for increasingly longer periods. Such layers can therefore be regarded only as protective measures for the vertical sealing layers against relatively coarse back-filling materials.

– Blockages can easily occur in layers with larger cavities as a result of rubble or soil residue which can fall in during back-filling and this can inhibit rapid water drainage.

Recommendations for the avoidance of defects

● If the whole of the back-filling material in the excavated site is to serve as a seepage layer (mixed filter), a type of material with sufficient permeability and stable filtering properties in relation to the surrounding soil should encase the drainage pipe. In most cases, graded gravel specified for concrete containing the lowest possible proportion of material which can be flushed out is a suitable material.

● Drainage walls erected directly in front of the external basement wall should take the form of dry-laid seepage materials made of porous concrete or rubble or drainage sheets made of polystyrene or polyethylene foam. In the case of fine-grained soils (silt), an additional filter layer (e.g. filter mat) should be laid between the drainage wall and the back-filling soil (see A 1.3.3). When using plastic drainage sheets their compressibility and durability, which are important factors determining their permeability characteristics, should be taken into account by employing suitable measures (e.g. greater thicknesses as the height of the drainage wall increases).

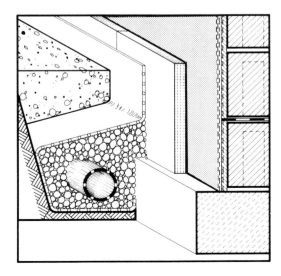

Drainage
Vertical seepage and filter layer

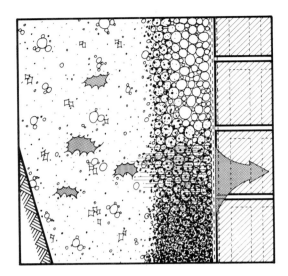

Even seepage layers that had suitable material structures showed signs of silting and drainage disorders, and these appeared as damp damage to the external basement walls after a short time.

Seepage layers made of gravel or 'filter concrete' in front of external basement walls were unable to filter away fine soil particles because they were insufficiently thick or because their structure was unsuitable.

Points for consideration

– Drainage layers or seepage layers in front of external basement walls have two functions to fulfil: they have to collect water and drain it vertically and they must be able to hold back fine particles of soil.

– The higher the proportion of continuous cavities within seepage layers, the greater their drainage capacity and drainage rate; however, this means that fine particles of soil can penetrate the cavities and silt the seepage system. A compromise must therefore be found between water permeability and the impermeability to fine soil particles.

– Slight penetration of fine soil particles is inevitable, even with layers whose construction has stable filtering characteristics. Thin seepage layers therefore tend to silt across their cross section.

– Perforated 'filter bricks' are not filter-proof from soil particles where the soil types to be filtered are very fine (silt, clay); because of their porosity and the presence of through joints and possible flaws, they allow a large proportion of fine particles to penetrate.

– If seepage/filter layers do not surround the whole of the area of the building subjected to water pressure, from the footings to ground level in the form of a closed system, or if damage or incomplete construction (especially in the case of projecting and receding structural components) have resulted in faults in the seepage wall, ground water drainage can no longer be guaranteed, at least in these areas, because of progressive silting and blockage.

Recommendations for the avoidance of defects

● Vertical drainage layers, which must surround the whole of the walls under pressure without interruption, should be sufficiently thick to ensure that an adequately large cross section will remain clear for water drainage after the inevitable silting of the outer areas. If the whole of the excavated site is not back-filled with water-permeable gravel with stable filtering characteristics right up to the drainage pipe, gravel layers with a minimum thickness of 500 mm should at least be built up in front of the external wall.

● If gravel layers are to remain permanently effective as drainage layers, their grain structure should have suitably stable filtering properties in relation to the surrounding soil. This can be done by building up the layers progressively to form a stage filter (e.g. from a 300–400 mm thick layer of coarse gravel (diameter 20–60 mm) and a 100–200 mm thick layer of sand (approx. diameter 0.2–4 mm) depending on the surrounding soil type) or a mixed filter (e.g. made from graded gravel specified for concrete work or grit).

● Filter layers (e.g. plastic or glass fibre mats) should in principle always be laid as a protection against silting if the grain structure of the drainage walls themselves (e.g. seepage bricks, gravel layers or plastic sheets) is not, or is only slightly, graded, and in any case in the presence of very finely grained soils. The filter layer must surround the drainage wall without gaps (overlap $\geq$ 100 mm at the joints) and without defective points.

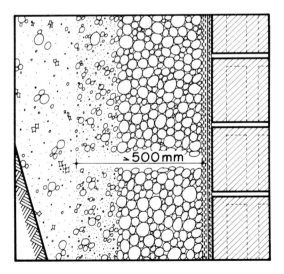

≥500mm

Text books and directives

Eggelsmann, Rudolf: Dränanleitung für Landbau, Ingenieurbau und Landschaftsbau. Verlag Wasser und Boden Axel Lindow & Co., Hamburg 1973

Muth, Wilfried: Dränung mit Dränrohren aus Ton. Fachverband Ziegelindustrie Baden-Württemberg, Stuttgart 1970.

Reichert, Hubert: Sperrschicht und Dichtschicht im Hochbau. Verlagsgesellschaft Rudolf Müller, Köln 1974.

Schultze, Edgar; Muhs, Heinz: Bodenuntersuchungen für Ingenieurbauten. 2. Auflage, Springer-Verlag, Berlin 1967.

Terzaghi, Karl; Peck, Ralph B.: Die Bodenmechanik in der Baupraxis. Springer-Verlag, Berlin 1961.

DIN 1180: Dränrohre aus Ton–Maße, Anforderungen, Prüfung. November 1971.

DIN 1185: Dränung–Regelung des Bodenwasser-Haushaltes durch Rohrdränung, Rohrlose Dränung und Unterbodenmelioration. Dezember 1973.

DIN 1187: Dränrohre aus PVC hart – Maße, Anforderungen, Prüfung. Januar 1971.

DIN 1986 – Blatt 2: Entwässerungsanlagen für Gebäude und Grundstücke – Bestimmung für die Ermittlung der lichten Weiten und Nennweiten für Rohrleitungen. August 1974.

DIN 4021 – Blatt 1: Baugrund–Erkundung durch Schürfe und Bohrungen sowie Entnahme von Proben, Aufschlüsse im Boden. Juli 1971.

DIN 4095: Baugrund – Dränung des Untergrundes zum Schutz von baulichen Anlagen, Planung und Ausführung. Dezember 1973.

DIN 18300 – VOB, Teil C: Erdarbeiten. August 1974.

DIN 18308 – VOB, Teil C: Dränarbeiten für landwirtschaftlich genutzte Flächen. Dezember 1973.

Scientific papers

Bellin, Kurt: Fortschritte der Dräntechnik aus der Sicht der Wasserforschung. In: Wasser und Boden, Heft 10/1973, Seite 322–325.

Bellin, Kurt: 10 Jahre Dränrohre aus Kunststoff. In: Wasser und Boden, Heft 11/1972, Seite 333–336.

Bruns, H.: Prüfung eines PVC-Dräns mit und ohne Kokosfilter im Dränkasten. In: Wasser und Boden, Heft 12/1971, Seite 350–353.

Knobloch, Hans: Die heikelste Stelle des Dränstrangs: die Wassereintrittsöffnung. In: Wasser und Boden, Heft 2/1969, Seite 34–36.

Muth, Wilfried: Dränung zum Schutz von Bauteilen im Erdreich. In: Forum-Fortbildung Bau, Forum-Verlag, Stuttgart 1977. Heft 8, Seite 115–127.

Muth, Wilfried: Abdichtung und Dränung am Bau, Schutz vor nichtdrückendem Wasser. In: Deutsche Bauzeitschrift (DBZ), Heft 1/1971, Seite 95–108.

Muth, Wilfried: Dränung am Bau mit Sickerplatten. In: Baumarkt, Heft 22/1975, Seite 580–588.

Muth, Wilfried: Dränung am Bau, Sickerschichten aus porösen Betonsteinen. In: Deutsche Bauzeitschrift (DBZ), Heft 5/1975, Seite 569–579.

Muth, Wilfried: Trockenlegen mit Polystyrol. In: Cunsulting + Technik, Heft 12/1976, Seite 34–36.

Muth, Wilfried: Dränplatten aus Styropor. Sonderdruck aus Deutsche Bauzeitschrift (DBZ), Heft 2/1973.

Muth, Wilfried: Schutzschichten vor erdberührten Wänden. In: Deutsche Bauzeitschrift (DBZ), Heft 8/1976, Seite 1025–1027.

Probst, Raimund: Außenwände im Boden – Dichtung und Dränung. In: Das Bauzentrum, Heft 2/1968, Seite 3–5.

Probst, Raimund: Baukonstruktive Risse. Analyse von Bauschäden. In: deutsche bauzeitung (db), Heft 2/1975, Seite 46–48.

Reichert, Hubert: Millionenschäden vermeiden. Sonderdruck aus deutsche bauzeitung (db), Heft 11/1975, Seite 51–53.

Rogier, Dietmar: Schäden und Mängel am Dränsystem. In: Forum-Fortbildung Bau, Forum-Verlag, Stuttgart 1977, Heft 8, Seite 68–75.

Saxen, A.; Karge, H.: Über die hydraulische Leistungsfähigkeit von Kunststoff-Dränrohren. In: Die Wasserwirtschaft, Heft 12/1968, Seite 362–364.

Wellenstein, Robert: Durchlässigkeitsbeiwert k (cm/s) des porösen Rohrmantels von Betonfilterrohren für laminare Sickerströmung. Sonderdruck aus Betonwerk + Fertigteil-Technik, Heft 1/1975.

Problem: Connecting structures

The various elements of a drainage system – pipes, drainage below basement slab, seepage and filter layer, main drainage system – and their correct interconnection should ensure that the whole drainage system is permanently operational.

Particular care should be devoted to designing and constructing transitional and connecting areas between the individual components as well as to the overall drainage system. This is particularly true because once the whole drainage system has been covered with soil it is very difficult to check and because its operation can only be checked indirectly (for example by means of flushing), so that possible faults and defects can no longer be clearly isolated. In this way, for example, a correctly constructed drainage system or vertical seepage layer of the correct size may not do their work if the necessary connection to the adjoining land drains has not been made correctly or if it has subsequently been interrupted during back-filling.

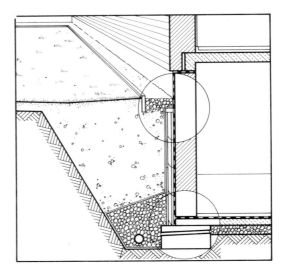

In all examples, defects took the form of flow disorders of varying severity in the drainage system, or the accumulation of water in the back-filled excavated site or under the floor slab as well as damp penetration of external walls, basement floor or their connecting areas associated with this. The extent and intensity of these defects varied according to the pressure and to the sealing measures employed.

The defects illustrated below which caused problems at various weak points centre around defective interconnection of pipes (pipe joints), connection of the vertical seepage layer to the adjoining land drainage (in the area of the footings) and/or in the area of the ground level as well as where the surface drainage connected with the land drainage.

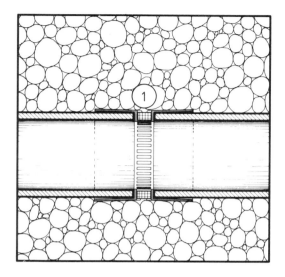

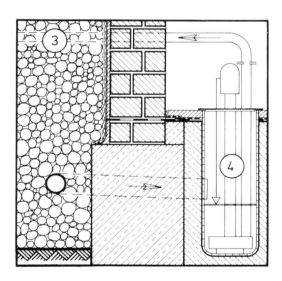

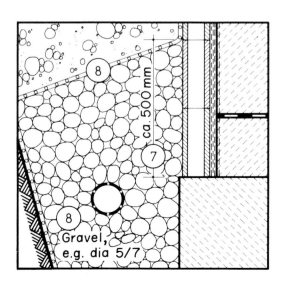

Gravel, e.g. dia 5/7

ca. 500 mm

1 In order to ensure a uniform joint gap, individual earthenware drain pipes should be connected with the aid of inserted rings and connecting sleeves, which at the same time fulfil a filtering role. Alternatively, drain pipes should be selected which are operational even if they do not have open joints, i.e. perforated or slotted plastic pipes and porous concrete pipes with connecting sleeves. The stable filtering properties of the pipes should, however, be proven (see A 2.1.2).

2 The drainage pipes must be laid as a closed system without interruptions on a firm, stable base, which should be checked before backfilling to ensure that the danger of subsequent pipe breakage is excluded. If necessary, the base of the seepage and filter layer surround should be consolidated with a layer of lean concrete (see A 2.1.2).

3 As a rule, drainage should only be laid when officially approved and a leak-proof connection to the main drainage system is possible without pumping equipment (see A 2.1.3).

4 If the use of pumping equipment in the drainage system is unavoidable, thorough, regular maintenance of the mechanical and electrical system should ensure that the pump and lifting equipment is constantly operational (if necessary install an emergency generator). The necessary collecting pit (sump) should be permanently accessible; it should therefore be inside the building whenever possible (see A 2.1.3).

5 Sections of pipe which are prone to become blocked should be protected by installing back-surge flaps (e.g. moulded discharge pieces made of plastic). The protection device must be easily accessible and it must be possible to check that it is constantly operational (see A 2.1.3).

6 In the case of connections to natural drainage systems, the drainage pipe should in principle discharge above the highest possible water level (see A 2.1.3).

7 The gravel layer surrounding the drain pipe must cover the footing of the draining wall with sufficient thickness (at least 500 mm from the main wall foundation) if rapid water dispersal is not guaranteed by other measures (e.g. seepage slabs lying flat on the projection of the foundations above the drainage pipe). Sufficient water permeability and sufficient filter-stability should be ensured by a graded composition of the gravel layer (see A 1.1.4).

8 Gravel layers with largely uniform grain structure (e.g. grain size diameter 5/7 or 8/10) must in any case be protected against silting by additional filter layers (filter mats). Moreover, in the case of fine surrounding soils or backfilled soil, it is advisable to include additional filter layers to surround the seepage layer – even where layers with graded grain sizes are used (see A 2.1.4).

Drainage

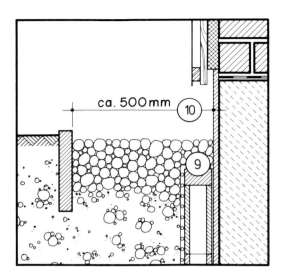

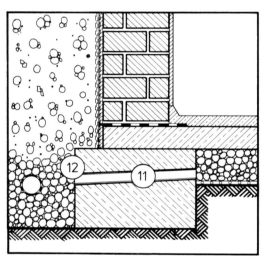

9 At ground level the upper terminations of vertical seepage layers should end about 200 mm below the level of the site and should be covered with material which is as pervious to water and as permeable as possible. In examples of drainage walls made of porous bricks, special protecting bricks made of the same porous material are suitable for this provided that their position is matched to the contours of the site (where the building is on a slope) or provided that they are stepped accordingly (see A 2.1.5).

10 A strip of gravel at least 500 mm wide and 200 mm deep (grain size approx. 32–63 mm) should be arranged in the area between the upper termination of the drainage wall and the ground level of the site (where the development permits) (see A 2.1.5).

11 Drainage arranged below the basement floor must be connected to the adjoining land drains. From this connection onwards, the nominal diameter of the drainage pipe should be increased in relation to the volume flow to be expected Drainage to the outside should best take the form of pipes (e.g. made of clay, concrete or plastic) cast in the foundations. These pipes should have sufficiently high pressure resistance. However, because of their number and arrangement within the cross section of the base, they should not endanger the stability of the building (see A 2.1.6).

12 Wherever possible, the connection from the underfloor drain should be installed in the upper section of the adjoining land drainage (see A 2.1.6).

Drainage
Connecting structures

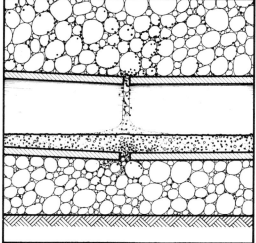

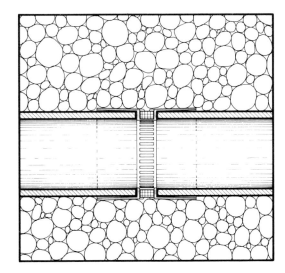

Flow disorders in the drainage system, with subsequent damp damage, particularly where the external wall of the basement connects with the base slab, occurred if large quantities of soil particles had been washed into the pipes through fairly large gaps in the pipe joints or through breaks in the pipe. The pipes concerned were mainly earthenware ones butted together without connecting sleeves, which had come apart by up to 50 mm and which were incorrectly laid or which had subsequently been broken.

Points for consideration

– In examples of earthenware drain pipes which do not have holes or slots in them and which do not have connecting sleeves, the quantity of water which they will collect depends entirely on the width of the joint. If, however, this width is greater than the grain size of the smallest particles of the seepage and filter material in the direct vicinity of the joint, silting up and blockage are to be expected – particularly where the gradient of the pipe is slight.

– In pipe joints without connecting sleeves, uniform correct width of the joint depends solely on the care with which the pipe-layer does its job: even if there are slight differences in the fall of the individual pipe sections there will be differences in the width of the joint along the line of the pipe.

– If the pipes are not laid on a stable base and if there are some cavities beneath the pipes, fairly large stones or an uneven or insufficiently compressed gravel bed, the possibility of pipe breakage as a result of subsequent loads (e.g. during back-filling) cannot be excluded.

Recommendations for the avoidance of defects

● In order to ensure a uniform joint width, individual earthenware drainage pipes should be connected with the aid of insert ring and connecting sleeves, which at the same time fulfil a filtering role. Alternatively, drainage pipes should be selected which are operational even if they do not have open joint, i.e. perforated or slotted plastic pipes and porous concrete pipes with connecting sleeves. The stable filtering properties of the pipes should, however, be proven.

● The drain pipes must be laid as a closed system without interruptions on a rigid, stable base, which should be checked before back-filling to ensure that the danger of subsequent pipe breakage is excluded. If necessary, the base of the seepage and filter surround should be consolidated with a layer of lean concrete.

Drainage
Connecting structures

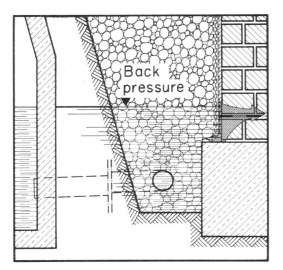

Water collected in the excavated site when there was no protection against back-surging at the point where the adjoining land drainage system connected with the main surface water drainage system. Water was fed from the main drainage system into the land drainage pipe, or in individual cases blockages were caused by faeces etc. which had been carried into the pipe where it was connected to a combined or waste water system. This resulted in damp damage, particularly where the walls joined the floor and in some examples even over the complete basement wall. If trust was placed in the operational effectiveness of the drainage and, for example, regular vertical wall sealing against temporary water pressure was only built up to a height of 500 mm, the back-surge of rising water which could occur was able to exceed this height and thus expose the wall to damp penetration. In other examples, the drainage water which accumulated was to be discharged to the main drainage system with the aid of pumps; however, the damage described above, as well as basement flooding, occurred as a result of defective pumps and lifting mechanism.

Points for consideration

– Natural drainage systems, such as open watercourses, as well as seepage systems, may result in fluctuating water or groundwater levels. This should be taken into account when setting the level at which the drainage is to be positioned.

– If main drainage systems (e.g. sewer pipes) are too small for the volume of drainage water, or if seepage systems are not connected to permeable layers which are capable of seepage, flow delays or disruptions can occur in the main drainage system as a result of heavy precipitation, rises in the groundwater or freezing conditions. These delays or disruptions can also cause back-surges in the drainage system or may render discharge of the drainage water impossible.

– If the drainage system is connected to waste or combined sewers, silting will also cause the drainage pipes to become blocked by faeces, etc. in most cases.

– If a land drain pipe cannot run to its junction with the main drainage system with a sufficient gradient, automatic lifting mechanism must be installed in order to feed the drainage water to the surface drainage system via pumps. In this case, disruptions (mechanical breakdown, failure of the power supply) may result in serious damp damage. Reliable operation of the system is heavily dependent on careful maintenance.

Recommendations for the avoidance of defects

● As a rule, drain pipes should only be laid either if they can discharge into the main drainage system (e.g. in open ditches) above the highest possible water level or if (e.g. in the case of rainwater drains) an officially approved connection to the main drainage system is possible above the back-surge level without the use of lifting equipment.

● If the use of lifting mechanism in the drainage system is unavoidable, thorough, regular maintenance of the mechanical and electrical system should ensure that the pump and lifting mechanism are constantly operational (e.g. by installing an emergency generator). The necessary collecting pit (sump) should be permanently accessible; it should therefore be inside the building whenever possible.

● Sections of pipe which are prone to back-surging should be protected by installing back-surge flaps (e.g. moulded discharge pieces made of plastic). The protection device must be easily accessible and it must be possible to check that it is constantly operational.

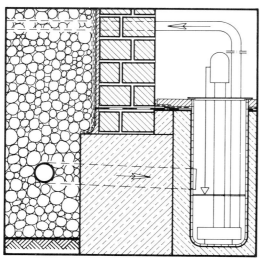

Drainage
Connecting structures

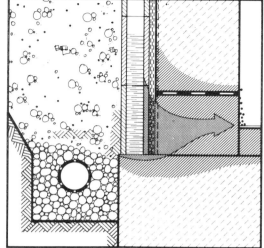

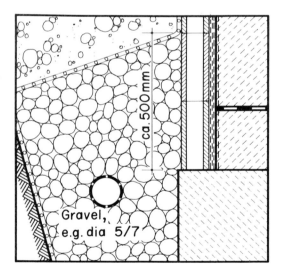

An incorrect base construction of an otherwise correctly built vertical seepage layer was frequently the reason why it could not fulfil its main functions of discharging the seepage water into the adjoining land drainage or why it did not operate adequately. Damage was most frequent in examples of drainage walls made of porous concrete blocks erected on the projection of the foundations, but it also occurred with other materials. In all examples studied, the permeable connection between the drainage wall and the drainage pipe was either not designed or constructed in advance or it subsequently had only a low flow capacity. The resultant damage took the form of damp penetration of the basement walls, particularly in lower parts of the structure.

Points for consideration

– If the water draining away through the vertical seepage layer cannot reach the drain pipe, it will accumulate in front of the external basement wall and exert pressure on it at least temporarily. If the wall is not sealed against this pressure (e.g. with bituminous coatings) damp penetration can be expected. The lack of a permeable connection between the vertical seepage layer and the drain pipe, or one which operates slowly, leads to this dangerous situation.

– These necessary 'seepage connections' show an inadequate flow capacity particularly when they are not sufficiently permeable as a result of their grain structure or pore distribution. Moreover, this can be reduced by water-borne soil particles to such an extent that in the end the connection is unable to fulfil its function.

Recommendations for the avoidance of defects

● The connection between the vertical seepage layer and the adjoining land drains should be constructed in such a way that water draining vertically through the drainage wall reaches the drain pipe unhindered. To do this, the gravel layer surrounding the drain pipe must cover the base of the seepage wall with sufficient thickness (at least 500 mm from the wall base) if a rapid water feed is not guaranteed by other measures (e.g. seepage slabs lying flat on the projection of the foundations above the drainage pipe). Sufficient permeability and sufficiently stable filtering characteristics should be ensured by a suitable grain composition of the gravel layer (see A 1.1.4).

● Gravels of largely uniform grain size (e.g. diameter 5/7 or 8/10) achieve adequate permeability, but they must be protected against blocking by filter layers (filtering mats).

● In examples of fine surrounding soil or back-filled material, it is advisable to include additional filter layers to surround the seepage material – even where layers with graded grain sizes are used.

Drainage
Connecting structures

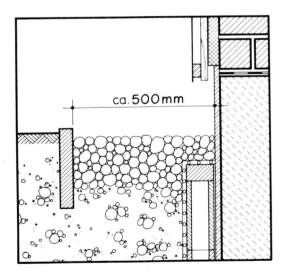

Parts of the upper areas of vertical seepage layers showed signs of damage if they terminated above ground level. In other examples, they were incorrectly constructed so that soil particles, stones or mortar residue which had fallen into the drainage wall resulted in blockage of the hollow cavities and in water accumulation and damp penetration of the external basement walls of varying severity.

Points for consideration

– The upper area of a drainage wall is particularly at risk if it extends above the ground into the open base area and if it is not protected against mechanical damage. Plastic drainage sheets in particular have a low resistance to this type of damage; in addition, they are not weather-proof (U.V./infrared radiation). Moreover, where the edge construction is visible, it is difficult to achieve a satisfactory appearance, particularly where the site contours vary.

– Seepage layers with large, continuous hollow cavities (e.g. drainage walls made of porous or hollow concrete blocks) may become blocked not only when they are built but also by soil particles (site waste) which can easily fall into them during back-filling if they are incorrectly constructed at the top. If this occurs, vertical water drainage to the drain pipe is either restricted or impossible, thus resulting in water accumulation inside the drainage wall with subsequent damp penetration of the external wall.

– Increased pressure by spray and rainwater is to be expected at the base of the external wall (foundation area) on sides subject to heavy rain.

Recommendations for the avoidance of defects

● Vertical seepage layers should terminate about 200 mm below the level of the site and should be covered with a layer of material which is as permeable as possible and which has the highest possible filtration capacity.

● In order to protect the upper connection of the seepage layer, in the case of drainage walls made of porous bricks, or blocks, special covering bricks made of the same seepage material are suitable provided that they follow the contours of the site (where the building is on a slope) or provided that they are stepped accordingly.

● A strip of gravel at least 500 mm wide and 200 mm in depth (grain size approx. 32–63 mm) should be arranged in the area between the upper termination of the seepage wall and the ground level of the site (where the development permits). Besides reducing the pressure on the base area caused by spray water, this ensures the most direct possible seepage of the rainwater from the foundation of the external wall to the drainage wall. At the same time, this prevents surface water carried to the foundation area from forming puddles (see B 2.2.5).

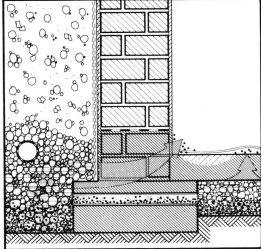

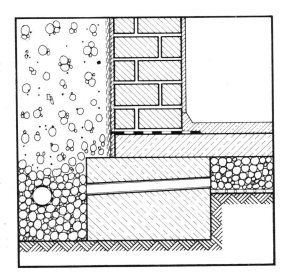

The accumulation of water beneath the basement floor slab and the penetration of damp into the interior of rooms where there were no sealing measures occurred if the surface drainage was not connected to the adjoining land drainage system. There were either no drainage facilities provided or they were rendered ineffective by incorrect design. In some examples, the land drainage was too high in relation to the surface drainage, so that connection was not possible.

Points for consideration

– Drainage is not complete and operational unless it is in a position not only to collect the water but also to discharge it, so that harmful accumulated water can collect. Generally, however, the space beneath the basement floor slab is surrounded on all sides by concrete strip foundations. Thus, drainage of the water which collects here is only possible through the foundation to an external drainage pipe of the land drainage system. This results in an increased load on the land drain and this must be taken into account when considering the pipe cross section (see A 1.1.3).

– The stability of the building may be endangered if the connecting method of the two drainage systems is not correct and, in the case of high water loading, if water can erode the base of the foundations or there is excessive weakening of the cross section of the supporting foundation. In all cases of doubt, therefore, a specialist engineer should be called in.

– If the drain pipe used to discharge the water is higher than the surface drain, or if drain pipes fitted in the foundations have an inadequate fall, or indeed if they fall in the wrong direction, there will be a relatively high risk of damp penetration of the basement floor because of slow and incomplete drainage and because of the prolonged period which the water remains underneath the floor.

Recommendations for the avoidance of defects

● Drainage arranged below the basement floor must be connected to the adjoining land drains. From this connection onwards, the nominal diameter of the drainage should be enlarged in relation to the volume flow to be expected. Drainage to the outside should preferably take the form of pipes (e.g. made of earthenware, concrete or plastic) taken through the foundations.

● The pipes should have a minimum diameter of 50 mm and a sufficiently high pressure resistance; they should not, however, endanger the rigidity of the building because of their number and because of the way in which they are arranged.

● The position of the drainage pipes must be correct in relation to the position of the adjoining land drainage and must be such that they can discharge the water reliably and completely. The drain pipe connections from below the basement floor slab should therefore, wherever possible, be installed in the deep section of the adjoining land drainage.

Drainage
Points of detail

Text books and directives

Muth, Wilfried: Dränung mit Dränrohren aus Ton. Fachverband Ziegelindustrie Baden-Württemberg. Stuttgart 1970.

Reichert, Hubert: Sperrschicht und Dichtschicht im Hochbau. Verlagsgesellschaft Rudolf Müller, Köln 1974.

DIN 1180: Dränrohre aus Ton – Maße, Anforderungen, Prüfung. November 1971.

DIN 1187: Dränrohre aus PVC hart – Maße, Anforderungen, Prüfung. Januar 1971.

DIN 1986 – Blatt 2: Entwässerungsanlagen für Gebäude und Grundstücke – Bestimmung für die Ermittlung der lichten Weiten und Nennweiten für Rohrleitungen. August 1974.

DIN 4095: Baugrund – Dränung des Untergrundes zum Schutz von baulichen Anlagen, Planung und Ausführung. Dezember 1973.

Scientific papers

Muth, Wilfried: Abdichtung und Dränung am Bau, Schutz vor nichtdrückendem Wasser. In: Deutsche Bauzeitschrift (DBZ), Heft 1/1971, Seite 95–108.

Muth, Wilfried: Dränung am Bau, Sickerschichten aus porösen Betonsteinen. In: Deutsche Bauzeitschrift (DBZ), Heft 5/1975, Seite 569–579.

Reichert, Hubert: Millionenschäden vermeiden. Sonderdruck aus deutsche bauzeitung (db), Heft 11/1975, Seite 51–53.

Rogier, Dietmar: Schäden und Mängel am Dränsystem. In: Forum-Fortbildung Bau, Forum-Verlag, Stuttgart 1977, Heft 8, Seite 68–75.

Schlünsen, D.: Kritische Bemerkungen zur Kunststoffrohrdränung und Beschreibung eines neuen Verfahrens der Tonrohrdränung mittels einer Kunststoffmuffenverbindung. In: Wasser und Boden, Heft 2/1969, Seite 32–34.

Problem: Sealing the external basement wall

Mortar group chart for brickwork

Mortar group	Non-hydraulic lime	Hydraulic lime	Cement	Sand
I		1		3
II	2		1	8
IIa	1		1	6
III			1	4

Mortar group chart for rendering

Mortar group	Non-hydraulic lime	Hydraulic lime	Cement	Sand
1 b		1		3
II	2		1	9–11
III			1	3

Damp protection is a characteristic function of the external basement wall. According to present methodology, there are a number of sealing materials and techniques for solving the problem of sealing external basement walls. In addition to bituminous asphalt coatings and priming compounds that have been in use for years, these include mainly mineral sealing materials, such as protective screeds and sealing renderings. Building paper, felt and plastic films for sealing buildings constitute a further group. In addition to these surface sealing materials, there is also water-repellent concrete, which is practically water-tight throughout its entire cross section.

Because of the changes in the uses for rooms with external walls below ground level which have been made on an increasing scale over the last few years, heavier demands are now made on damp protection. Whilst for storing provisions, a high level of humidity in the room and thus a high moisture level in the walls is desirable in many cases, and indeed sometimes necessary, more intensive use of these rooms by individuals, or the storage of moisture-sensitive goods requires a sufficiently dry room atmosphere with impermeable external walls. In some examples, an additional moisture load caused by the use of the room must also be taken into account.

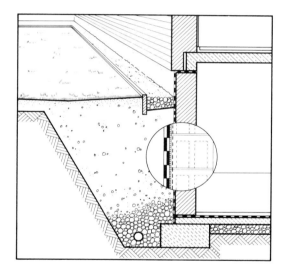

By far the most common type of damp damage was to basement walls. One of the main factors to emerge from the survey is the fact that, besides underestimating the water pressure placed on the external wall as a result of water penetration beneath the floor, the sealing effect of different types of seal is frequently very much over-estimated (see O 1.1.0). For this reason, considerable importance must be attached to analysing the inefficiency of types of seal under practical building conditions and to deducing from this the areas in which a particular method can be used, allowing for a reasonable safety margin.

Below is a description of the types of seal used in appreciable quantities in building construction together with their weaknesses and typical faults. Recommendations for preventing these faults are then derived.

External basement wall

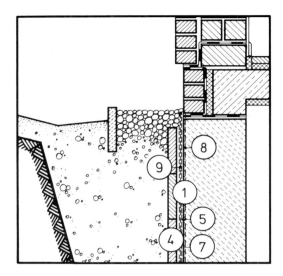

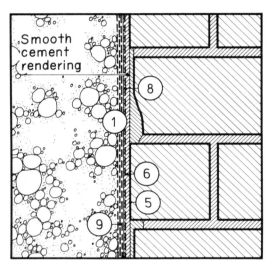

1 Bituminous coatings, water-proof rendering, sealing compounds or water-proof concrete (for which there is no evidence of its liability to cracking based on thickness) should only be used to seal external basement walls when the only pressure consists of soil moisture and non-accumulating seepage water. If effective drainage is provided, these sealing methods can also be employed in pressure caused by the temporary accumulation of seepage water (see B 1.1.2, B 1.1.5 and B 1.1.8).

2 External basement walls should only be sealed by the adhesion of two layers of building paper if the only pressure to which they are subjected is the temporary accumulation of seepage water. If effective drainage is provided, building paper can also be used where the pressure is seepage water accumulating over long periods (see B 1.1.13).

3 Seals against seepage water and groundwater which accumulates over a long time must be made in the form of tanking of at least three layers of building paper or untearable felt or from water-proof concrete made by specialist firms (see B 1.1.13 and B 1.1.15).

4 Surface seals made of bituminous coatings, water-proof rendering, sealing compounds and building papers or felt must be protected with a material layer capable of resisting damage after they have been produced and particularly when the working space is back-filled (see B 1.1.2, B 1.1.5, B 1.1.8, B 1.1.12 and B 1.1.13).

5 The wall surface to be sealed with bituminous coatings must be made of a uniform, solid material without flaws and irregularities, which is free of impurities and loose particles and which has a rough surface (see B 1.1.3).

6 Before applying cold bituminous asphalt coatings, brickwork should be primed with a highly adhesive smoothing render (MG III) to cover all irregularities. This smoothing render should not be applied until settlement of the external walls caused by the load of the basement ceiling slab has occurred – after a period of at least three months (see B 1.1.3).

7 Where possible, bituminous asphalt coatings should be applied directly to concrete without a priming layer. Thus, a plumb surface must be obtained by careful shuttering and the concrete must have a textured finish (see B 1.1.3).

8 A priming coat must be applied to the dry substratum once it has been prepared for the bituminous coatings. If, under exceptional circumstances, wall surface which is still damp is to be sealed, only asphalt should be used (see B 1.1.3).

9 Bituminous seals for external basement walls must consist of at least two hot-laid or three cold-laid coats applied carefully and evenly to the dried priming coat. If a stronger sealing layer is required, priming pastes can also be used and should be applied in two layers either hot or cold to the priming coat by a trowel or a float (see B 1.1.4).

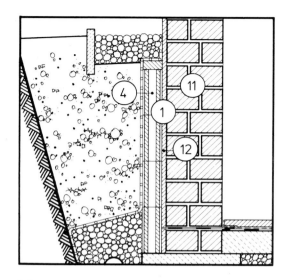

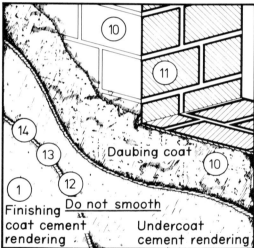

Daubing coat

Finishing coat cement rendering

Do not smooth

Undercoat cement rendering

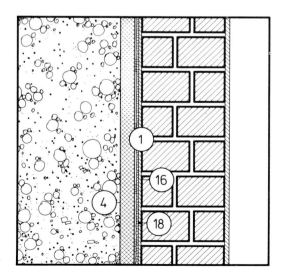

10 Water-proof rendering should be applied only to a wall surface made of a uniform, solid material without flaws and large irregularities. Efflorescence and dirt, as well as ridges in the surface of the concrete, should be removed and open joints in the brickwork should be filled. The dry substratum should be sufficiently dampened. If a coarse rendering coat (MG III with 7 mm diameter sand) is required on smooth or absorbent substrata, this should be applied one day before final rendering (see B 1.1.6).

11 Where possible, the water-proof rendering should not be applied until settlement of the external basement walls caused by the load of the ceiling slab and the upper storeys has occurred – i.e. after a period of three months (see B 1.1.6).

12 Water-proof concrete should be made from cement mixed in ratio of 1:2 to 1:3 with sand of grain size 0–3 mm and with approx. 20 wt. % of fine particles between 0 and 0.25 mm, together with a sealing compound (see B 1.1.7).

13 The rendering should be applied to the carefully prepared substratum in at least two layers and in a minimum thickness of 20 mm. It should be applied either by throwing it by hand or with a plastering machine in such a way that adequate key and compression are produced. The second layer of rendering should be applied to this sufficiently firm but still damp render base. If this is not possible, a coarse rendered coat should be applied to the dried lower coat of render (see B 1.1.7).

14 There should be a difference in the strength of the outer and inner layers of render. The inner render coat should therefore contain more binding agent (ratio 1:2) than the outer render coat (ratio 1:3) (see B 1.1.7).

15 The rendering work should not be carried out in direct sunlight or strong wind. If necessary, the fresh rendering coat should be covered for at least forty-eight hours and kept damp. Work should be planned in such a way that continuous surfaces can be completed without interruptions. If it is necessary to connect finished work, the individual layers should be overlapped by about 150 mm (see B 1.1.7).

16 The wall surface to be sealed with a sealing compound should consist of a uniform, solid material without flaws and irregularities, should be free from impurities and loose particles and should have a rough surface (see B 1.1.9).

17 The substratum for the sealing compound should not have any visible surface cracks, nor should there be any visible cracks at points where it connects with other parts of the structure. It should not suffer any settlement which would lead to breaks or cracks in the wall cross section (see B 1.1.9).

18 Before applying sealing compounds, brickwork should be treated with a highly adhesive smoothing render (MG III) to cover all irregularities. This smoothing render, and particularly the sealing compounds, should not be applied until the settlement of the external walls caused by the load of the basement ceiling slab and upper storeys has occurred – after at least three months (see B 1.1.9).

External basement wall

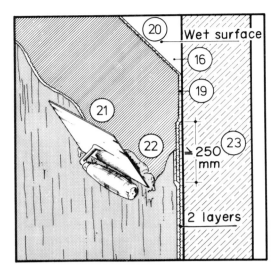

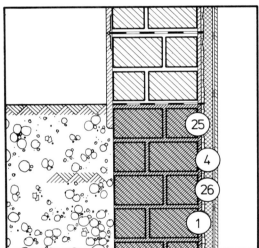

19 Where possible, the sealing rendering should be applied directly to concrete without the use of a priming coat. Thus, a plumb surface must be obtained by careful shuttering. If the surface is too smooth, adhesion straps should be used (see B 1.1.9).

20 When applying the sealing compounds, the substratum must be saturated with water, but not so that it glistens with moisture. It should therefore be sufficiently dampened at least one hour before applying the compound in order to ensure that the water penetrates deep into the substratum. Smooth wall surfaces which are not highly absorbent should be dampened at least twelve hours before commencing work (see B 1.1.9).

21 The sealing compound should be mixed with water to form the finished render coat immediately before starting work. It should be prepared with stirring equipment and should be applied to a carefully prepared and dampened substratum by throwing it and spreading it with a trowel or float. It is also possible to apply it with a spraying machine (see B 1.1.10).

22 The sealing compound should be applied in at least two layers, with the second layer applied as soon as possible after the first coat has set and reached a sufficient degree of strength. Wherever possible the minimum quantities specified by the manufacturer should be exceeded. If the lower layer has already set, the subsequent layer should not be applied until it has hardened fully and until its surface has been prepared in the appropriate manner (see B 1.1.10).

23 Sealing work using sealing compound should be planned in such a way that connected surfaces can be finished without interruptions. If it is necessary to connect finishing work, the individual layers should be overlapped by at least 250 mm (see B 1.1.10).

24 Sealing compounds should not be prepared in direct sunlight, rain or strong wind. Freshly finished wall surfaces should be covered and kept damp for at least forty-eight hours (see B 1.1.10).

25 In examples of new buildings subjected to soil moisture and non-accumulating seepage water, the external basement wall should only be sealed internally with water-proof render or sealing compounds when it is impossible to provide an effective external seal and when harmful effects on parts of the structure are counteracted adequately (see B 1.1.11).

26 The air in basement rooms whose external walls are sealed internally will have a higher humidity level and the insides of the walls should be covered with thermal insulation layers and, where possible, with a surface finish which can store moisture (e.g., anhydrous plaster). These finishes also act as protective layers (see B 1.1.11).

27 If sections of the external basement walls which are not accessible from the outside have to be sealed internally, the whole wall should be sealed wherever possible, and not just the section concerned. At least one metre of partition walls which connect with this wall should also be sealed (see B 1.1.12).

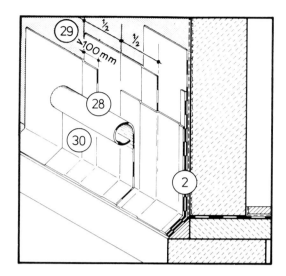

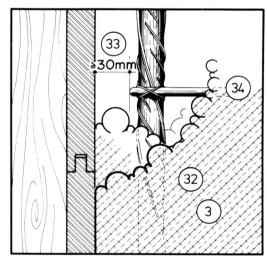

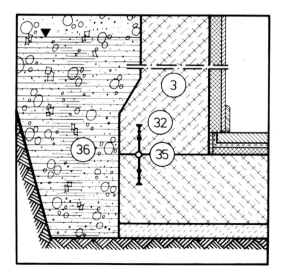

28 Two layers of building paper or untearable felt should be adhered to the whole surface in such a way that the longitudinal joints are overlapped by half the width of the paper and the lateral joints are overlapped by at least 300 mm. The building paper or felt should overlap at the joints by at least 100 mm (see B 1.1.14).

29 The substratum for building paper or felt must be level, dry and free from loose particles. In examples of concrete walls, all burrs should be removed and edges and channels should be rounded off. The use of a smoothing render should be avoided where possible. Brickwork should be coated with a highly adhesive coat of smoothing render (MG III) to level out all irregularities (see B 1.1.14).

30 Factory produced sealing layers (excluding those with inserts which are liable to decay), plastic films for sealing buildings and adhesive compounds are suitable for use as sealing materials (see B 1.1.14).

31 Sealing systems consisting of building paper or felt should only be constructed in dry weather and at outside air temperatures of at least + 5°C (see B 1.1.14).

32 Water-proof concrete should be made from a dense aggregate with a maximum grain size of 16 or 32 mm and with a graded sieving sequence, plus at least 350 kg/m³ of cement and an adequate proportion of fine grains. The cement water value must be below 0.6. Concrete sealing agents may be added, provided that they have been proved to be suitable (B 1.1.17).

33 In the case of water-proof structures, the minimum overlap and the distance between the reinforcing rods including the tie bars should be 30 mm and should be at least 5 mm larger than the maximum grain size of the additive (see B 1.1.17).

34 The water-proof concrete should be poured into the shuttering in such a way that separation is avoided and a complete seal is achieved. No auxiliary products which will remain in the concrete or which will be removed once the setting process has begun (racking wires) should be used (see B 1.1.17).

35 The work should be planned in such a way that interruptions do not occur. Joints in the work should be avoided wherever possible by, if necessary, adding retarding agents. The position and construction of unavoidable joints in the work should be planned, for example by fitting suitable gaskets, and by cleaning and preparing the surface of the concrete (see B 1.1.17).

36 Structures made of water-proof concrete must be protected against drying out and frost. They should be kept damp for at least twenty-one days and, if necessary, should be covered with thermal insulation mats. Water-proof concrete walls which have been removed from the shuttering should be filled with soil immediately afterwards where possible. If not, the surface of the wall should be protected with several layers of bituminous asphalt coats (see B 1.1.17).

37 Because of the large number of factors which must be taken into consideration in producing water-proof concrete, this work should be entrusted to specialist firms (see B 1.1.17).

External basement wall
Sealing the external basement wall

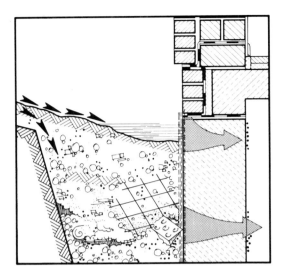

The external basement walls of residential buildings are most frequently sealed with bituminous asphalt coatings. In by far the majority of examples, the damage that occurred was to walls protected with this type of seal. In many cases, damage even occurred to walls where the finishes were perfect, especially if the building was erected on a slope or in cohesive soil without effective drainage measures and if it was exposed to accumulating seepage water. In these examples, moisture had penetrated large areas of the external basement walls and damage appeared in the internal plaster and wallpaper, and indeed water also ran into the building through joints in the brickwork or through cracks in the wall. Damp damage also appeared, even where there was little exposure to soil moisture or seepage water, if the bituminous seals had been exposed to prolonged weathering or if they had suffered impact loads during construction or back-filling.

Points for consideration

- Even if they are applied in several layers, bituminous coatings are still not very thick. Their effectiveness therefore depends mainly on their ability to fill any cavities on the surface of the wall and lies less in producing a waterproof skin. Whilst in examples of slight water exposure this largely eliminates capillary water absorption by the wall surface, water which is present under pressure and in relatively large quantities may penetrate the wall through the pores of the coating skin, particularly in the case of painted bituminous finishes. Then it will reach the inside of the wall as a result of capillary action or by penetrating hollow cavities.

- Bituminous finishes may sustain damage as a result of fairly prolonged exposure to the weather. U.V. radiation renders the coating brittle. Heat, particularly in the form of sunlight directed on the black surface, softens the bituminous compound. Rain may wash off a finish which has been painted on the wall.

- Because it is so thin, a seal made of bituminous finish will become damaged even by relatively slight mechanical pressures (friction, impact) so that flaws will appear.

- Similarly, the finishing film is unable to cover and bridge completely even slight cracks in the substrate (render, supporting wall cross section), especially at low temperatures.

Recommendations for the avoidance of defects

● External basement walls should only be sealed with bituminous coatings when they are exposed only to soil moisture and non-accumulating seepage water.

● If effective drainage is provided, bituminous coatings can also be used where the wall is exposed to brief accumulations of seepage water.

● After they have been constructed, and particularly during back-filling of the site, seals made of bituminous coatings should be protected by a layer that is resistant to mechanical pressures (see C 1.2.6).

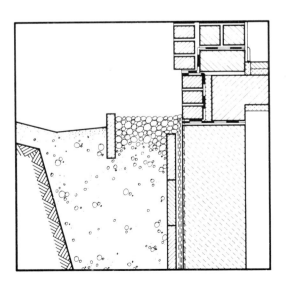

External basement wall
Sealing the external basement wall

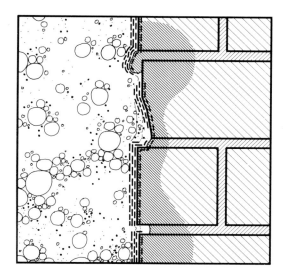

A large proportion of the damage to external basement walls sealed with bituminous finishes, which took the form either of surface damp, fungus growth and damage to wallpaper or layers of plaster, or which were visible, for example, on the internal plaster above joints in the brickwork, were largely due to an inadequate substratum for the sealing coatings. Bituminous finishes which had been applied direct to the brickwork had flaked or cracked above joints which had receded or over projecting pieces of mortar or stone edges. In examples of concrete surfaces, the coatings were unable to cover ridges in the shuttered wall, exposed gravel and projecting reinforcing wires and thus had flaws at these points. If the bituminous coatings had been applied to render, flaws could be attributed to points where the render had chipped away from the substratum, or to cracks in the render which were either irregular or which followed the joints in the brickwork. In addition, there were cases where the bituminous coating had peeled away from wall surfaces that had not been cleaned.

A further cause of imperfect seals is cracks in the supporting wall in cross section, which damaged both the rendering and the coating film.

Points for consideration

– The thin coating film, particularly where it is applied cold in solution or painted form, is not able to bridge completely irregularities in the substrate, such as the porous surface of aggregate blocks, unfilled joints in brickwork or shuttering ridges in the case of concrete walls; it tears above hollow cavities or becomes thin at the edges of brush strokes. A coating that has been sprayed on is better.

– Loose particles (dust, soil residue), efflorescence or shuttering oil residue prevent the coating from penetrating the pores of the walling material and form a separating layer and this has an adverse effect on the adhesion of the coating film. Since certain coatings are relatively viscous because of the film thickness to be produced, their moistening effect and their ability to penetrate the substratum are inadequate. Because of this, priming coats are applied. These are very fluid and are therefore able to penetrate the brickwork or concrete wall and to form a binding layer for the subsequent coatings. If the substratum is damp, a solven-based priming coat has difficulty in penetrating the water filled pores of the walling material. The moisture is sealed in by the coating and so possible exposure to sunlight may result in blistering and peeling of the coating.

– Bituminous painted finishes can be applied to damp substrata, but this increases the drying time. If exposed to rainfall, they may be washed away before they have dried. Frost may also cause damage.

– A substratum made of materials with varying degrees of absorbency leads to coatings of varying thicknesses, for example at points where concrete and brickwork meet as well as between brick and mortar joints. Because of its thinness and brittleness at low temperatures, the coating film is unable to bridge fine or wide cracks which appear in the substratum but will crack itself. In terms of susceptibility to cracking, bituminous coatings can be compared with rigid types of sealing finishes.

External basement wall
Sealing the external basement wall

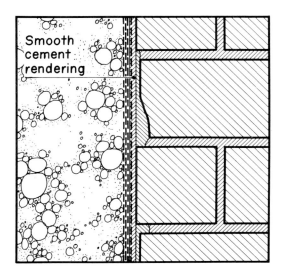

Recommendations for the avoidance of defects

● The wall surface to be sealed with bituminous coatings must consist of a solid, uniform material without flaws and irregularities, must be free of impurities and loose particles, and must have a rough surface texture.

● Before applying cold bituminous finishes, brickwork should be prepared with a highly adhesive smoothing render (MG III) to even out all irregularities. This smoothing render should not be applied until at least three months have elapsed in order to ensure that the settlement caused by the load of the basement ceiling slab has occurred. Smoothing render is unnecessary in the case of hot bituminous asphalt coatings.

● Wherever possible, bituminous finishes should be applied to concrete directly without priming render. For this reason, an even surface must be produced by a smooth, rigid shuttering and by sufficient compression of the concrete when poured.

● A priming coat should be applied to a prepared, dry substratum. If under exceptional circumstances, the wall surface to be sealed is still damp, only painted finishes should be used.

External basement wall
Sealing the external basement wall

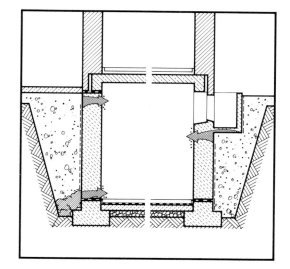

In the majority of examples of damage covered by the survey, the measures used to seal the external basement walls with bituminous coatings took the form of one or two coats applied cold to rendered brickwork or to concrete. Accordingly, the finishes failed to seal up pores in the wall surface and were prone to holes and flaws, so that internal wall surfaces became damp. In addition, there were examples of damage that were caused by fairly large wall surfaces not covered by sealing coatings, especially those under projecting balcony slabs, basement stairs and light shafts. In some cases, the seal had not been continued as far as the lower horizontal wall damp-proof course or had not been joined to the base area.

Points for consideration

– For the effective sealing of building structures against soil moisture, it is prescribed that the priming coat should be three cold or two hot bituminous asphalt coatings. If applied correctly, the average coating thickness to be achieved in this way is about 1 or 3 mm. Every reduction in the number of layers and in the quantities used reduces the thickness of the coating. This also reduces its ability to fill the pores of the walling material and to produce an effective seal.

– Priming coats serve to produce an adhesive bond between the substratum and the covering layers. They thus do not have the function of producing a seal. The type of priming coat used (emulsion or solution) depends on the moisture content of the substratum and on the type of covering coats used.

– The application of cold liquid coating appears initially to involve less work. However, it entails more operations and must be completely dry before the next layer is applied. Finally, because they are thinner, they have less resistance than hot layers.

– The undersides of structural components which project from the wall, such as basement stairs, light shafts and balcony slabs, are areas which are difficult to reach and to check in carrying out sealing work.

Recommendations for the avoidance of defects

● Bituminous coatings must be applied without breaks to the whole of the wall surface that is below ground level and connect with the horizontal wall and floor sealing system or with the base construction.

● The external basement walls should be sealed by applying at least two hot or three cold liquid layers of bituminous material to the dried priming coat. If the sealing layer is required to have greater resistance, filler compounds can also be used. They can be applied either hot or cold to the priming coat with a trowel or with a float.

● If cold liquid finishes are used, special attention should be given to applying all the layers evenly. Each layer must be fully dried before applying the next coat.

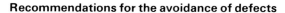

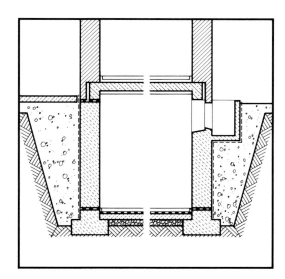

External basement wall
Sealing the external basement wall

Damp damage to external basement walls with water-proof rendering was particularly in evidence when the buildings were erected in soils not very permeable or on a slope or in groundwater. This meant that the external basement walls were exposed to water under pressure. External basement walls erected against a slope were damp across their whole surface area. In examples of brickwork, water had frequently penetrated at points on the surface and there were muddy brown run marks from joints in the brickwork. Water also ran into the basement through cracks in the brickwork.

If external basement walls which had already been sealed were left for a long time before being back-filled, or if the site was back-filled by machine with excavated soil containing stones and building debris, the outer sealing layer became damaged, resulting in visible damp on the inside.

Points for consideration

- As the test reports that many manufacturers of sealing agents for water-proof rendering enclose with their product information sheets show, mortar prepared with such products is largely impervious to water, even with high hydrostatic pressure. The results of empirical building damage research show, however, that in many cases, water-proof renderings applied to the structure according to the manufacturer's instructions did not fulfil their function, particularly when exposed temporarily to accumulated water.

- Because of this, very high demands are made in terms of producing the rendering mortar and of applying the water-proof render (see B 1.1.7). These demands are frequently not met, particularly in the case of relatively small building projects.

- Water-proof renderings are very brittle in use because they are composed of mineral building materials and are therefore unable to balance out movements in the substratum at fractures and at the angled profiles edges of materials without damage. Instead, they crack at these points. The same applies to internal stresses in the render layer itself. In examples of new buildings – expecially those made of brick and non-reinforced concrete – cracking cannot be excluded.

- Impact loads when the excavated site is still open, or during mechanical back-filling with stones or sharp edged building debris (concrete waste, pieces of steel), may damage the water-proof concrete in such a way that its sealing function is impaired.

- Flaws in a sealing layer are very serious in the case of water under pressure, since fairly large quantities of water can penetrate through them and can then reach the inside under pressure through hollow cavities and flaws in the substratum (cracks, hollow joints). Compared with this, the quantity of water transported by capillary action is smaller.

Recommendations for the avoidance of defects

● External basement walls should only be sealed with water-proof rendering when the walls are exposed to soil moisture and to non-accumulating seepage water.

● If effective drainage is provided, water-proof rendering can be used to seal the walls where they are exposed to seepage water that accumulates temporarily.

● After they are produced, seals made of water-proof render must be protected by a layer which is resistant to mechanical loads when the excavated site is back-filled (see B 1.2.6).

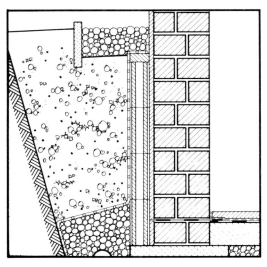

External basement wall
Sealing the external basement wall

Cracks and peeling in water-proof rendering, which also resulted in damp damage on the inside of the external basement wall, could usually be attributed to the composition and inadequate preparation of the render.

If cracks had appeared in the wall in cross section, these continued into the rendering and they also appeared at points where various parts of the structure joined (steel or concrete support – brickwork) or at changes in the wall materials (from the concrete to the section of wall built up to it). Thin render had fine cracks along joints in the brickwork. Hairline cracks were observed in water-proof render applied to very uneven brickwork. The render peeled where the concrete walls were smooth and also when the wall surface was soiled, for example by soil, dust or cement slurry.

Points for consideration

– The protective effect of a water-proof render depends among other things on its adhesion to the base wall and on the fact that it is free of cracks and flaws. These properties are significantly affected by the composition of the base wall.

– Good adhesion depends on the roughness and absorbency of the substratum. A rough base (dense concrete blocks) promotes adhesion particularly by keying with the render base. In examples of less textured but absorbent substrata (brick) adhesion is produced by the cementing action of the cement lime which penetrates the pores. Water-proof concrete can therefore only be applied to a smooth substratum with low absorbency (concrete) if the surface of the substratum is first roughened (sand-blasting, or by applying rough daubed render base). When applying the render to a highly absorbent substratum (e.g. aerated concrete) too much of the mixing water may be removed from the render base, so that it becomes parched, i.e. so that it cannot harden completely. The removal of water causes the cement to set more quickly, thus reducing its curing time. An absorbent substratum must therefore be wetted. In the case of a highly absorbent substratum, a rough render coat is first necessary to reduce the absorbency.

– Unevenly thick coats of render, which occur when the render base is uneven, also cause variations in the setting and drying process. This produces inherent stresses in the rendered surface which may lead to cracking.

– Because of its brittleness, mineral render is not able to withstand movements without damage in the substratum such as those which occur over cracks, at joints and at points where various structural components or materials connect.

Recommendations for the avoidance of defects

● The surface of the wall to be sealed with water-proof render must be of a firm, uniform material without flaws (open joints, cracks) and large irregularities. Efflorescence and soiling, as well as ridges on the surfaces of concrete components should be removed and open joints in the brickwork should be filled.

● A dry substratum should be sufficiently wetted. If the render is to be applied to a smooth or highly absorbent substratum, a coarse layer of render (MG III with sand diameter 7 mm) is necessary and should be applied one day before the plastering work is carried out.

● Wherever possible, the water-proof render should not be applied until at least three months have elapsed in order to ensure that building settlement caused by a sufficient loading of the basement ceiling slab and of the upper storeys has occurred.

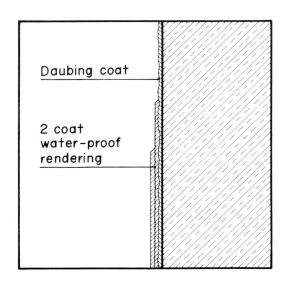

Daubing coat

2 coat
water-proof
rendering

External basement wall
Sealing the external basement wall

In many examples, inadequately produced water-proof renderings did not fulfil their sealing function, so that damp damage appeared on external basement walls as a result of soil moisture or the temporary accumulation of water. Damp over large areas of the walls could be attributed to a porous and water absorbent rendering mix to inadequate rendering thicknesses of only a few millimetres or to reticular hairline cracks in the render.

Irregular damp patches occurred in external walls where the thickness of the water-proof render varied widely. This rendering was generally applied in one layer to a very uneven substratum, usually brickwork. Where render produced with water-proof additives was applied in several layers, the top layer of render peeled.

Individual damp spots could be attributed to the following. Some areas, particularly at the base of the wall, as well as in inaccessible places, such as under projecting light shafts, staircases, etc., were not covered with water-proof render, the type of render below the horizontal wall damp-proof course changed to MG II (external wall render) or individual sections of the work were formed without overlapping the layers of render.

Points for consideration

- The sealing effect of a water-proof rendering depends on the composition of the mortar, on the addition of a sealing agent, on careful preparation and application, and on producing a rendered surface of sufficient thickness.

- Below ground level, the render is exposed to moisture continuously. The ingredients of the mortar must therefore be water-proof and only cements which set hydraulically are suitable.

- The type of additive has a significant effect on the pore volume and thus on the density of the render. A good pore structure and good working characteristics are achieved using sand with a grain size of 0–3 mm which is constantly graded and in which the proportion of particles with a grain size of 0–0.25 mm amounts to 15–25 wt.%.

- Mortar sealing agents have only a reinforcing sealing effect. The imperative prerequisite for the effectiveness of the render is a physically optimum cement mortar with a small pore structure which is thus impermeable. Sealing agents available today achieve their effect by closing pores since the ingredients in them swell or coat the walls of the pores with materials which render them hydrophobic. They are either added to the mixing water on site (fluid sealing agents) or to the cement or additive-cement mixture (powdered sealing agents). An effective mortar is not produced until the individual ingredients have been intimately mixed by a mechanical mixer for longer than three minutes.

- These strict requirements in terms of material selection and preparation of the water-proof render are often not definitely fulfilled on smaller sites. Ready mixed mortar would be useful in such cases.

- Rendered coats less than 20 mm thick generally do not achieve the desired sealing effect. This is particularly disadvantageous in the case of an uneven substratum. Water-proof rendering 20 mm thick can normally not be applied in one operation. Excessively large variations in the thickness of the render will, however, be avoided if the required thickness of at least 20 mm is achieved in two layers of 10 mm each. In doing this, each render layer must be applied evenly and without cavities over the whole of the substratum.

External basement wall
Sealing the external basement wall

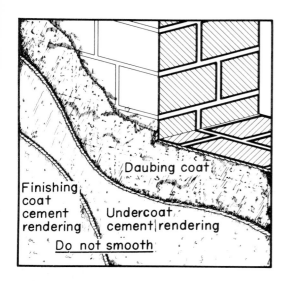

Finishing coat cement rendering

Daubing coat

Undercoat cement rendering

Do not smooth

– If the top layer of render contains more binding agent and is therefore harder than the bottom layer, shrinkage stresses may cause the top layer to crack and peel. If the top layer of render is applied to a layer of render which has already set fully and dried, or if this layer has not been textured sufficiently, adhesion will be adversely affected and this may also lead to peeling. This is particularly true of water-proof renders. Prolonged abrasion of the render will increase the amount of cement lime on the surface and this will result in shrinkage cracks on drying.

– The water necessary to produce settings may be removed from fresh render by a highly absorbent substratum, or as a result of increased evaporation at the surface caused by high air temperatures, direct sunlight and wind. This reduces the strength and particularly the density of the render.

Recommendations for the avoidance of defects

● Water-proof render must be applied to the whole surface of the external basement wall below ground level and must connect with the horizontal wall and floor seals and with the base construction.

● Water-proof render should be prepared from cement mortar mixed in the ratio of 1:2 to 1:3 (MG III) with sand with a grain size of 0–3 mm with a finest grain size of 0–0.25 mm representing approximately 20 wt.%, plus a sealing agent.

● The mortar should be thrown on to a carefully prepared substratum by hand or with a plastering machine in such a way that a sufficient key and seal are produced.

● The water-proof rendering should be applied in at least two layers and with a minimum thickness of 20 mm. The second layer of render should be applied to the first layer when it is sufficiently firm, but whilst it is still damp. If this is not possible, a rough coating of render should be applied to the dried first layer (see B 1.1.6).

● There should be a difference in the strength of the first and second layers of render. The mixture of the first layer of render should therefore contain more binding agent (mixing ratio 1:2) than the second layer (mixing ratio 1:3).

● The rendering work should be organised in such a way that adjacent surfaces can be finished without interruption. If it is necessary to join parts of surfaces, the individual layers should be overlapped by about 150 mm.

● Rendering work should not be carried out in direct sunlight or high wind. If necessary, the fresh rendered surfaces should be covered and kept damp for a period of at least forty-eight hours.

External basement wall
Sealing the external basement wall

Even perfectly executed seals to external basement walls involving sealing compounds did not prevent damage, especially when the building was constructed in cohesive soils that were not very permeable or if they were constructed on a slope and if because of this the soil was damp for prolonged periods or the basement walls were exposed to groundwater.

In individual cases, patches of damp in the external basement walls could be attributed to damage to the sealing compounds caused by external factors during building or when back-filling the excavated site. In brick walls, joints stood out in the internal render or water ran out of the joints in the brickwork leaving behind muddy brown deposits. In concrete walls, the points at which the water emerged were usually pockets of unmixed gravel or parts that had been inadequately sealed, as well as flaws caused by reinforcing wires, spacers, etc. Damp damage was particularly predominant if there were cracks in the external walls as a result of varying degrees of building settlement, and the sealing compounds were affected by these cracks.

Points for consideration

– As laboratory tests have shown, sealing compounds are largely waterproof even when they are exposed to water under pressure. Nevertheless, damp damage occurred in many new buildings, especially when the external basement walls were temporarily exposed to water under pressure.

– Sealing compounds consist mainly of materials of mineral origin and are applied in very thin layers to the external basement walls. The compounds must therefore be applied with special care.

– Movements in the substratum, which occur mainly in the period immediately after construction, for example at cracks, joints or places where layers of material connect, result in fractures in the sealing compounds. In the case of newly built structures made of brick or non-reinforced concrete, the processes which produce such cracks cannot be excluded in practice.

– Impact loads on the sealing compounds, such as those which occur during construction when the excavated site is still open, as well as when the site is back-filled with stones, building debris, etc., may damage the thin layer and thus result in leakage.

– Where there is exposure to soil moisture, individual flaws may allow water to penetrate the wall in cross section through capillary action, although the feed of water will be small in terms of the size of the flaws. Where there is exposure to accumulated seepage or groundwater, however, large quantities of water can penetrate the wall in cross section through small individual breaks in the seal and these will reach the inside of the wall through hollow cavities, open joints, etc. and will cause damp damage.

Recommendations for the avoidance of defects

● External basement walls should only be sealed with sealing compounds where they are exposed to soil moisture and non-accumulating seepage water.

● If effective drainage is provided, sealing compounds can also be used if the pressure consists of seepage water which accumulates temporarily.

● After they have been formed, as well as during back-filling of the excavated site, seals made of sealing compounds should be protected by a layer which is resistant to mechanical pressures (see B 1.2.6).

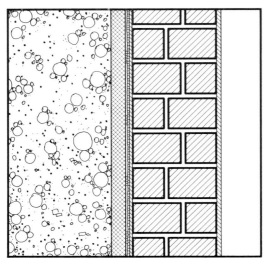

External basement wall
Sealing the external basement wall

Faults in the sealing compounds applied to external basement walls, which usually resulted in damp damage to basement rooms, were sometimes caused by faults in the substratum.

Cracks appeared in the sealing compounds above wide cracks in the cross section of walls which were usually made of brickwork or non-reinforced concrete, as well as at points where different parts of the structure connected and at areas where different wall materials (concrete, brickwork) joined.

Where sealing compounds were applied directly to brickwork surfaces, joints in the brickwork appeared in certain examples in the form of hairline cracks. If the joints were hollow, or if there were other flaws and irregularities in the brickwork, the compound layer peeled away at these points, or became very thin at the edges of bricks or at ridges, especially when it had been applied with a brush. In the case of sealing compounds applied to concrete surfaces, faults appeared above pockets of gravel, ridges in the wall surface or reinforcing wires. The sealing compounds were hollow or peeled away from the substratum, together with the bottom coat of rendering in some cases, particularly on concrete surfaces, but also on brickwork soiled by earth, dust, etc.

Points for consideration

– Sealing compounds are applied to a wall surface as a sealing base in a layer only a few millimetres thick. In order to ensure effective sealing, which depends on sufficient thickness, good adhesion and permanent freedom from flaws and cracks, very high demands are placed on the compound substratum.

– Even slight irregularities in the substratum, such as shuttering ridges in concrete or incompletely filled joints in brickwork, cause the compound layer to peel away or at the very least result in uneven layer thickness, which will in turn lead to flaws or joint damage because of uneven setting. A sufficiently good and even surface can be given to brick walls by applying a coat of smoothing render (MG III).

– Because of the thinness and the brittleness of mineral sealing compounds they are unable to withstand without damage even slight movements in the substratum, such as those which may appear at cracks, joints or at points where different materials and structural components connect, as a result of irregular settling and shrinking and creeping processes. On such substrata, sealing compounds are unsuitable for use as a sealing layer.

– The absorbency of the substratum has a very improtant effect on the density and the adhesion of the sealing compounds. Where the substratum is too dry or too absorbent, the necessary mixing water for complete hydration of the first layer of compound and thus for achieving the required strength and water-proofness may be removed from the compound. This can only be avoided reliably if the substratum is saturated with water. A shining wet surface, however, which appears when there is excessive water, will impair the formation of an adequate bond between the sealing compound and the substratum.

– Soiling, excessive cement lime, organic components (shuttering oil) and soft material will impair perfect adhesion of the sealing compound to the substratum.

External basement wall
Sealing the external basement wall

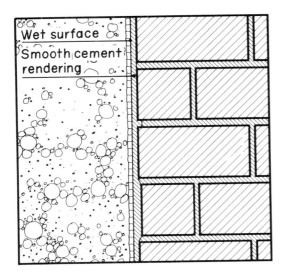

– Good adhesion is achieved when the substratum is rough and firm, since the lower layer of sealing compound is able to key with the substratum. If the substratum is smooth (high grade reinforced concrete) adhesion can be improved by using adhesion strips (according to manufacturer's instructions).

Recommendations for the avoidance of defects

● The surface of the substratum and points where it connects should consist of a firm, uniform material without flaws and irregularities and must be free of impurities and loose particles. It should also have a textured surface.

● The surface of the substrate and points where it connects with adjacent structural components should have no visible cracks and they should not undergo any subsequent changes in shape or length which would result in cracking or in the widening of cracks in the wall in cross section.

● Before applying the sealing compounds, brickwork should be provided with a highly adhesive smoothing render (MG III) in order to level out all irregularities. This smoothing render, and particularly the sealing compounds, should not be applied until a period of at least three months has elapsed in order to ensure that building settlement in the external basement walls caused by the weight of the basement ceiling slab has occurred.

● Wherever possible, sealing compounds should always be applied directly to concrete without a priming coat of render. For this reason, a level surface must be produced by good, firm shuttering and by adequate compression of the concrete poured. If the surface is too smooth, adhesion strips should be used.

● When applying the sealing compound, the substratum must be saturated with water, but should not be so wet that it glistens. It should thus be wetted sufficiently at least one hour before applying the compound so that the water is able to penetrate deep into the substratum. Smooth wall surfaces which are not very absorbent should be wetted at least twelve hours before work is started.

External basement wall
Sealing the external basement wall

In a large number of cases, damp damage to external basement walls could be attributed to sealing compounds which had not been applied correctly.

In most cases, the layer of compound was too thin or uneven, since it was applied with only one stroke of the brush. In particular, on unrendered brick walls, the sealing compounds did not cover completely or were porous. This resulted in damp damage to large surfaces of the external basement walls.

Isolated patches of damp on some parts of the walls could be attributed to the cracked or hollow top layer of the sealing compounds or to incorrectly made overlapping joints between sections of wall which had been sealed at different times. Damage also occurred if not all the surface was covered with sealing compound. Such flaws occurred frequently at the footings of the external basement walls and in inaccessible places under basement stairs, projecting light shafts and balcony slabs.

Points for consideration

– The sealing effect of a sealing compound is based on having a layer of mortar with a dense texture. The mortar consists of cement and finely ground quartz sands plus chemically or physically active additives which constrict pores (substances which swell) or which have a water-repelling effect (rendering materials hydrophobic).

– Unlike water-proof renders, sealing compounds are supplied as mortar which is ready mixed in the factory. Faults in the mixture are therefore largely excluded on site provided that the mixing water is added immediately prior to commencing work in accordance with the manufacturer's instructions. Mixing must ensure that the water is distributed evenly throughout the product, and this is best achieved using mechanical equipment.

– Sealing compounds are applied in thin layers of only a few millimetres. Because of this, even small irregularities reduce the thickness and thus have an adverse effect. It is easier to avoid variations in thickness if the sealing compounds are applied in at least two coats and if a certain quantity is applied per square metre in each coat. The covering capacity quoted by the manufacturer is usually a minimum and greater quantities should be used wherever possible.

– Uneven thicknesses, flaws and porous compound layers are easily produced and in many cases these cannot be overcome by applying several layers. There is less risk of such defects if the sealing compounds are applied by skilled personnel with a trowel or float or with a spraying device.

– Interruptions in the work were often the reason for leaks at points where parts of surfaces joined. Because of this, layers of compound which have already dried require renewing and careful treatment of the substratum. This can be avoided if the next layer of compound is applied as soon as possible after the previous layer has hardened initially and if the work is not interrupted.

– In the case of absorbent substratum, or where evaporations is increased by high air temperatures, direct sunlight or wind, the water needed to make the compound set may be removed from freshly applied sealing compounds. This has an adverse effect on texture and strength.

External basement wall
Sealing the external basement wall

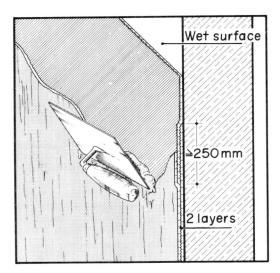

Recommendations for the avoidance of defects

● Sealing compounds must be applied without holes to the whole surface of the wall which is below ground level and must connect with the horizontal wall and floor seal or to the base construction.

● Mortar should be prepared from the sealing compound by adding water immediately before work is started using mechanical equipment and should be applied to a carefully prepared and wetted surface (see B 1.1.9) by throwing it on and spreading it with a trowel or float. It is also possible to apply the compound with a spraying machine.

● At least two coatings of sealing compound should be applied, the second layer being applied as soon as possible after the first layer has hardened initially and has become sufficiently strong. Where possible, more compound should be used than the minimum quantities specified by the manufacturer. If the lower layer has already set the following layer may not be applied until the former has hardened fully and its surface has been treated in the appropriate manner.

● The work should be organised so that adjacent surfaces can be finished without interruptions. If it is necessary to connect parts of a surface, the individual layers should be overlapped by at least 250 mm.

● Sealing compounds should not be applied in direct sunlight, rain or high wind. Freshly finished wall surfaces should be covered and kept moist for a period of at least forty-eight hours.

External basement wall
Sealing the external basement wall

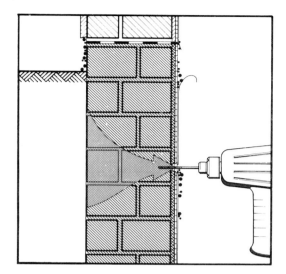

External basement walls are not often sealed internally with water-proof rendering or sealing compounds. Some of the examples of damage covered by the survey occurred in inadequately sealed external basement walls which had subsequently been sealed internally. The damage which was observed corresponded largely to the damage which occurred in water-proof rendering and sealing compounds applied to the outside of walls. Where walls were exposed to water under pressure (accumulated and groundwater) damp penetration occurred and water emerged from cracks or flaws in the substratum, such as hollow joints in the brickwork or at points where adjacent structural components met and at expansion joints in the concrete walls.

Points for consideration

– When the sealing coat is applied to the inside of the external basement wall, the whole cross section of the wall is permanently damp. This has several effects and may possibly lead to damage.

Only materials permanently resistant to damp and water should be used in the external basement wall and, in the area of the foundations, the materials must be frost resistant to a depth of at least 1 m. Reinforced concrete should only be used if it is certain that no water is present which would damage the steel and concrete. In any case, stricter demands should be made on the concrete covering of the reinforcement and in terms of the impermeability of the concrete. In such instances it is preferable to build the basement from water-proof concrete because of the small additional cost involved.

– Over the course of time the water which penetrates makes the whole cross section damp and this moisture rises up through the wall by capillary action. Here it is able to place pressure on parts of the structure that are not directly concerned but which must definitely be protected against damp (basement ceiling, external wall of ground floor). Above ground level the water will affect the base surface, where it will evaporate leaving behind materials which it was carrying in solution, or may cause peeling (for example in frosty conditions). The height of the area of the vertical wall prone to damp will depend on the capillary conductivity of the materials of which the wall is made, the water feed and the drying capacity of the surface. By placing horizontal damp-proof courses in the wall's cross section beneath the basement ceiling slab it is possible to interrupt this passage of water (see B 2.2.3). This type of protection is impossible for internal walls. Thus, depending on the type of wall material and the evaporation capacity, these walls will become damp to varying depths and must therefore be included in the internal sealing measures used.

– The thermal insulation value of the damp wall in cross section drops considerably when damp has completely penetrated it. For example, in the case of an external basement wall made of common bricks, the thermal insulation value is about one third of that when the wall is dry. This results in greater heat losses and might result in condensation on the top surface of the wall. The dense structure of the mineral sealing layers also prevents temporary storage of condensation in the surface of the wall. Because of this, additional thermal insulation measures on the inside of the wall are necessary, and also layers which are capable of storing moisture (see B 1.2.4).

External basement wall
Sealing the external basement wall

– There are cases where, because of site conditions, neighbouring buildings, or in places such as underneath stairs, for example, which are inaccessible from the outside, the execution of effective external seals is impossible. This is also true when renovating damp walls. In these cases, internal seals are a constructive way of counteracting possible harmful effects.

– One of the advantages of internal seals compared with all types of external seal is the fact that they are easily accessible and that they can be directly checked. Leakages can be pinpointed and specifically repaired.

– Because flaws can be pinpointed and repaired cheaply compared with external seals, an internal seal made of water-proof render or sealing compound can also be used where the wall is exposed to accumulated water provided that cracking processes in the substratum (resulting from shrinkage or creeping) can be largely excluded.

– See B 1.1.5 and B 1.1.8 for further details.

Recommendations for the avoidance of defects

● Sealing the external basement wall internally with water-proof render or sealing compound is a constructive form of damp protection. Because of the possibility of adverse effects associated with this type of seal, it should only be used if an effective external seal is impossible and if adverse effects on parts of the construction can be adequately counteracted.

● The external basement walls of new buildings should only be sealed internally with water-proof render or sealing compound if the walls are exposed only to soil moisture and non-accumulating seepage water.

● If effective drainage is provided, and if cracking processes in the substratum are largely excluded, internal sealing can be used even if the walls are subjected to seepage water which accumulates temporarily.

● The external basement walls of rooms whose internal atmosphere has a high moisture content (relative air humidity) should be covered internally with thermal insulation layers and if possible with a surface finish which can store moisture (e.g. anhydrous plaster) (see B 1.2.4).

● In the base area or in the area below the basement ceiling, permanently effective horizontal seals must be incorporated into the external basement wall.

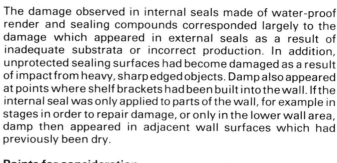

External basement wall
Sealing the external basement wall

The damage observed in internal seals made of water-proof render and sealing compounds corresponded largely to the damage which appeared in external seals as a result of inadequate substrata or incorrect production. In addition, unprotected sealing surfaces had become damaged as a result of impact from heavy, sharp edged objects. Damp also appeared at points where shelf brackets had been built into the wall. If the internal seal was only applied to parts of the wall, for example in stages in order to repair damage, or only in the lower wall area, damp then appeared in adjacent wall surfaces which had previously been dry.

Points for consideration

– If a seal is placed on the inside of the wall, for example at points which allow insufficient access to provide an external seal, the water which penetrates makes the wall damp; and this damp spreads throughout the whole cross section of the wall, since evaporation in the inside of the room is largely impossible. If the internal and external seals do not overlap by a sufficient amount damp may even appear on the inside of the wall or in partition walls connected to these areas which are some distance away from the place where the water entered from the outside. The width of the wall area that must be sealed both on the inside and on the outside depends on the capillary conductivity of the wall material, and the degree to which the internal seal is vapour-proof as well as the climate in the basement rooms, and on the exposure to water. Although where the exposure consists of soil moisture and non-accumulating seepage water the quantity of moisture carried under capillary action is relatively small, where the wall is exposed to accumulated water, large quantities are able to penetrate relatively large sections of wall.

– If they are subjected to impact loads from sharp-edged objects, mineral render or compound layers may be pierced and they will thus lose their sealing effect. Any penetration which occurs at a later date (e.g. caused by a wall plug) will have the same effect. An additional rendered coating (15 mm), for example, will provide protection against impact loads, as will a thermal insulation layer which has been plastered on to the surface. Compared with external sealing layers covered with soil, the internal seal has the advantage of making it possible to pinpoint faults and to repair them specifically.

– Please see B 1.1.6 or B 1.1.9 for details of the substratum composition necessary for achieving an effective seal using water-proof render or sealing compounds. In respect of material selection and preparation of the water-proof render or sealing compounds, please refer to B.1.1.7 or B 1.1.10.

Recommendations for the avoidance of defects

● If internal seals are to be applied to sections of external basement wall which cannot be reached from the outside, it is necessary to seal not only the appropriate section of wall but also, wherever possible, the whole wall. Partition walls which connect with the wall should also be sealed to a width of at least one metre.

● The internal seal must be connected to the horizontal damp-proof course of the external basement wall in the base area or below the basement ceiling slab and to the seal of the basement floor in such a way that moisture bridges cannot arise.

● The seal should be protected by an internal coat of render 15 mm thick or by some other coatings which are resistant to mechanical loads (e.g. a thermal insulation layer).

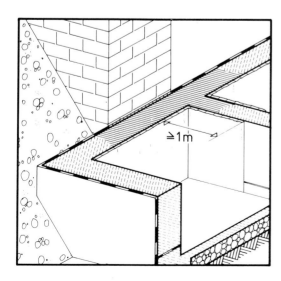

≧1 m

External basement wall
Sealing the external basement wall

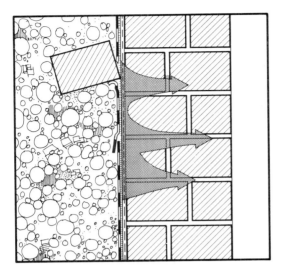

Seals to external basement walls which involve building papers or plastic films were unable to prevent damage in many cases where walls were exposed to water under pressure. Where these seals were applied to the wall surface in a single layer, both the lateral and the longitudinal joints had come away or were not water-proof, as were the connections to the basement floor seals which were taken through the basement wall. In this way water was able to penetrate the wall in cross section via the wall surface and/or through the lower section of the wall. In some cases, only the vertical external surfaces of the basement walls were sealed without overlapping them with the lower horizontal wall damp-proof course so that water was able to enter the basement and the rising wall in cross section from the foundations. Damp damage occurred as a result of individual flaws which could be attributed to impact loads which took place during construction or during back-filling. The fact that building papers became loose and peeled away also resulted in leaks. Some cases of water damage occurred where multi-layer tanking was used. These were caused by the fact that water ran over the top edge of the seal and ran back behind the building papers which had been inadequately adhered to the substratum.

Points for consideration

- In the cases of damage surveyed as part of this work, only a relatively small number concerned water-pressure tanking. This is clearly due partly to the technical standards, guidelines and practical experience relating to this type of pressure and partly to the care exercised by trained personnel in forming tanking constructions. The majority of examples of damage covered concerned single and double layer building papers without backing layers which were exposed to accumulated water.

- The surface of building papers and plastic films is practically water-proof. The waterproof characteristics of a wall seal thus depend mainly on the seal produced by the lateral and longitudinal seams of the courses. Faults when adhering the materials are hardly to be avoided under normal building site conditions. Therefore, in the case of single layers of building paper, the points at which the seams are adhered represent weaknesses. In contrast, in the case of double layers of building paper over the whole wall surface with offset seams, the risk of continuous faults is considerably less. Where seals can only be repaired at a later date with difficulty and at great expense it is particularly important to keep the risk of damage as low as possible.

- Producing a seal made of sealing courses which only reaches up to a certain wall height and providing a type of seal which is suitable for a lighter pressure above it, for example in the form of several coatings of bituminous material (see B 1.1.2), requires a reliable and precise knowledge of the water pressure, and particularly of the maximum accumulated water level. Because of the inevitable uncertainties in determining or estimating the maximum water level, the seal should be made 0.5 m higher than necessary.

- External basement wall seals consisting of building papers which do not have a water-proof connection with the seal for the basement floor are inadequate where the excavated site is exposed to water under pressure, since the water can run in behind the sealing layers. In such examples, all parts of the building or structure which have to be kept dry must be completely surrounded by sealing layers.

External basement wall
Sealing the external basement wall

– Until the excavated site is back-filled, the seal applied to the wall surface is exposed to the weather. High air temperatures and especially direct sunlight acting on the black surface results in the risk that the sealed wall surface will become very hot. A slight reduction in peak temperatures can, for example be achieved by applying a coat of lime. The most effective method, however, is to shade the seals or to cover them with protective layers. Adhesion alone is not able to hold a double layer of sealing material to a vertical wall surface, mainly because of the limited strength of the adhesive layer. It is not possible to use adhesives of sufficient strength because of the complications associated with using them (high temperatures). Because of this, the only way of achieving a suitably strong seal is to attach the upper edge of the building papers mechanically.

– Wall seals which are left unprotected against conditions on the building site may be subjected to mechanical damage (friction, impact) and develop holes.

The risk of this occurring is particularly high when back-filling mechanically with the excavated soil which has possibly been mixed with building debris. Moreover, when tipping the soil from a relatively high position and in large quantities, the sealing layer which has been adhered to the wall surface may become displaced and may even be torn away from the attachment at its top edge.

Recommendations for the avoidance of defects

● External basement wall seals using two layers of adhesively fixed building paper or plastic films should only be used where the wall is exposed to seepage water which accumulates temporarily.

● If effective drainage is provided, double layer seals can also be used where the wall is exposed to seepage water which accumulates for prolonged periods.

● Where the walls have prolonged exposure to accumulated seepage water or groundwater, the external basement walls should be sealed with at least three layers.

● After it has been made, and particularly when the excavated site is back-filled, a seal made of building paper must be protected by a layer which is resistant to mechanical damage (see B 1.3.6).

● The seal applied should cover all the structural components to be protected, which normally consist of the external basement walls and the basement floor, without gaps and in two layers. It should cover the external basement walls up to about 200 mm below ground level and should be attached mechanically at this point.

External basement wall
Sealing the external basement wall

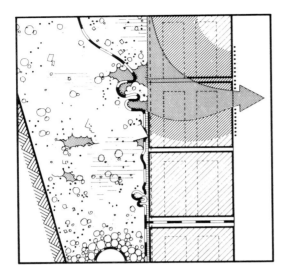

Leaks, and in some examples destruction, of wall seals made of building paper or plastic films could be attributed to incorrect construction. Insufficiently overlapped or inadequately adhered or welded lateral or longitudinal seams between the paper resulted in leaks where the seal consisted of a single layer. If the whole surface of the paper was not fixed by adhesive to the substrate or if a layer which had been adhered on in wet or frosty conditions had become loose, any water which penetrated ran behind the seal and caused damp damage over large areas of the wall. In cases where the building paper was not attached at the top by mechanical means, or where this was not done properly, it tore away and sagged in folds on the wall surface. If the upper edge was not secured against seepage water, rainwater seeping in front of the wall surface got underneath the sealing skin.

Points for consideration

- The effectiveness of seals using building paper or untearable felts depends mainly on the careful construction of the overlaps at the longitudinal and lateral seams and on ensuring that the whole surface of the paper is adhered to the substratum. Any impurities, water or air which become trapped behind the paper impaired this complete adhesion. In cold conditions, the wetting properties of the adhesive layer will be lost, whilst a film of water present when the hot adhesive compound is applied will act as a separating layer when it evaporates. This can be expected in particular if it rains or if it is cold when the adhesive is applied. The risk of leaks is reduced considerably if several layers of building paper are applied carefully and are overlapped by half of the width of the paper below, since faults in one layer cannot have any direct harmful effect because of the complete coating of adhesive between the layers.

- A substratum which has excessively large irregularities, for example brickwork which has not been rendered, as well as a substratum which is damp or soiled will not ensure complete adhesion of the paper. Artificial drying of the surfaces (e.g. with a gas burner) is generally not adequate because the effect of the heat is not deep enough. Emulsion based priming coats apply water to the substratum. This can be avoided by using solvent-based cold liquid finishes.

- Complete adhesion of the building paper to the substratum prevents the possibility of water which has penetrated through leaks in the sealing layer spreading over the surface and causing extensive damage.

- Increased exposure to seepage water can be expected in the upper area of the wall, firstly because of the film of water which runs down the façade in heavy rain and secondly because of the surface water to which the base is exposed. If the building paper does not run under the wall surface or if the top edge of the seal is not covered by the horizontal wall sealing layer of the base area, water may run in behind the sealing layer.

External basement wall
Sealing the external basement wall

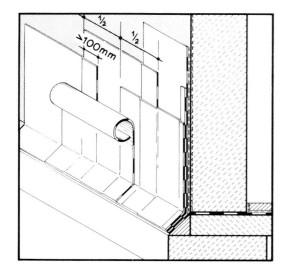

Recommendations for the avoidance of defects

● Building papers should be applied in two layers with the adhesive applied to the whole of their surface area. The longitudinal seams of the paper layers should be overlapped by half of the width of the courses and the lateral seams should be overlapped by at least 300 mm. At the seams, the paper layers must be covered by at least 100 mm. The substratum for building paper must be level, dry and free of loose particles. In the case of concrete walls, all shuttering ridges should be removed and edges and fillets rounded off. The use of smoothing render should be avoided where possible. Brickwork should be covered with a highly adhesive smoothing render (MG III) to smooth out all irregularities.

● The following are suitable as sealing materials:
– factory-produced building paper strips, except those with inserts which are liable to decay.

– untearable roofing felt without mineral finish.

– plastic films for sealing buildings.

Filled adhesive compounds (softening point according to prevalent climatic conditions $\geq 100°C$) should be used as adhesive layers.

● Seals made of building papers and plastic films should only be applied by skilled personnel in dry weather and at external air temperatures of at least +5°C.

● The upper edge of the seal, which should be continued until about 200 mm below ground level, must be protected against seepage water (e.g. by introducing it into a bed joint in the brickwork) and must be attached mechanically (e.g. by a 100–150 mm metal strip screwed into the wall).

External basement wall
Sealing the external basement wall

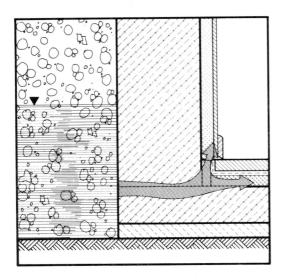

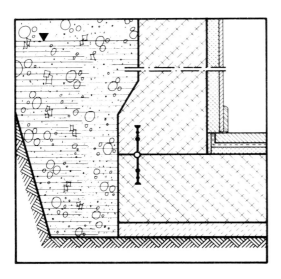

Damp damage occurred in external basement walls made of concrete in the presence of varying water pressure in the soil. Large areas of the walls became damp if the concrete had a porous structure or if pockets of gravel were present. Water penetrated the basement or damp affected the wall in areas along joints in the work (concreting lift joints) or in joints between the elements of prefabricated walls as well as in the area of reinforcing wires and spacers inadequately covered. Damp damage and water penetration were particularly in evidence at wide cracks in the basement walls.

Points for consideration

– The degree to which water-proof concrete is water-tight depends on a texture which is particularly dense and contains few cavities. For this reason, extremely heavy demands are made in terms of the selection of materials especially where the additive and water cement value are concerned, and in terms of the mixing, application and subsequent treatment of the cement. Concrete sealing agents can only improve the water-proof characteristics of cement which has been prepared in the best possible way and which is therefore already largely water-proof.

– Cracks in walls represent leaks. There is a great danger of cracking when foundations settle in an irregular manner and when concrete walls dry out too quickly. Leaks can also be expected in the joints between the elements of prefabricated basement walls. For this reason, particularly heavy requirements are made on basement walls made of water-proof concrete in terms of measures to prevent cracking.

Recommendations for the avoidance of defects

● External basement walls made of water-proof concrete, for which there is no evidence of their limitation of cracking width, are suitable as a seal only when they are exposed to soil moisture and non-accumulating seepage water.

● If effective drainage is provided, a water-proof concrete wall made in this way can also be used if it is exposed to seepage water which accumulates for short periods.

● External basement walls made of water-proof concrete, which are exposed to accumulating seepage water or groundwater for prolonged periods may only be constructed by specialist companies in order to keep cracking width down to less than 0.1 mm.

External basement wall
Sealing the external basement wall

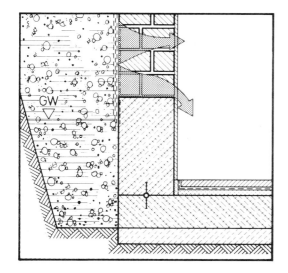

A large proportion of the damage to external basement walls made of water-proof concrete could be attributed to insufficient reinforcement and defective construction. Damp in basement rooms caused by wide cracks occurred mainly in water-proof concrete walls which were not reinforced or which were inadequately reinforced. There were also leaks between prefabricated concrete sections and those which were made on site. All these parts were subjected to varying loads (columns, wall surfaces). In examples where a water-proof tanking was erected up to a certain height and where brickwork was then placed on top, the joint between the concrete and the brickwork, and sometimes the wall section too, leaked.

Points for consideration

– Constructing a water-proof concrete wall to a certain height requires a reliable and precise understanding of the type of water exposure, particularly of the maximum height of the water table. As a rule, the maximum water table level is not known reliably, particularly since, for example, when the building is on a slope, it may be raised by local conditions (adjacent buildings) or indeed by the building itself (accumulation of groundwater, additional surface water or water flowing down a slope towards the building).

– Cracks can be expected particularly in non-reinforced concrete and in concrete which has been inadequately reinforced. There can be various reasons for this, such as varying degrees of settling in the site, ground pressure connected with inadequate reinforcement, inherent stress caused by shrinkage and temperature conditions.

– Joints between various elements are not crack-free, even if they have been filled with mortar. This is true where a site erected concrete wall is connected to a prefabricated column or to one which was made at an earlier date. Cracks in edge layers caused by an inadequate bond and by shrinkage, as well as by differences in load and settling in the case of separate foundations, are unavoidable. The degree of freedom from cracks is increased if the basement is constructed as a continuous, rigid tanking made of water-proof concrete, built in one operation if possible.

– The importance of cracks in terms of their having an adverse effect on water-proofing depends mainly on the width of the cracks themselves. In the case of cracks over 0.2 mm wide, we can expect the passage of increasing amounts of free water under pressure in additon to water carried by capillary action. Such cracks lead large quantities of accumulated or groundwater into the basement rooms. However, free water finds it increasingly difficult to penetrate cracks less than 0.2 mm wide. Thus, water enters these cracks mainly by capillary action and the quantity transported is small.

– Generally, reinforced concrete, which is a composite material of concrete and steel, is made in such sizes that if tensile stress occurs the action of the concrete is impaired This implies that if appropriate use is made of the steel stresses, cracks will appear in the concrete around the area of the stress. In the case of reinforced concrete parts which must be water-proof (capillary water, water under pressure), tensile stresses in the concrete must be reduced by selecting a suitable bearing system, taking the tensile strength of the concrete into account. In addition, stresses caused by varying or irregular temperatures and shrinkage must also be taken into account. In doing this, it is sometimes necessary to increase the proportion of steel reinforcement and in particular, its distribution and binding effect (as many small pieces of steel as possible, placed close together) must be correct.

External basement wall
Sealing the external basement wall

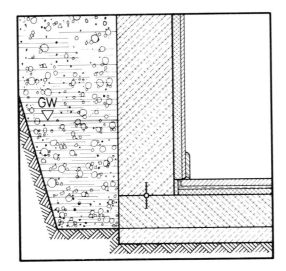

Recommendations for the avoidance of defects

● If water-proof concrete is to be used as a seal against groundwater or prolonged accumulation of seepage water, those external surfaces of the building which are in the ground (external basement walls and basement floor) must be built in the form of tanking.

● The size of water-proof tanking should be selected with a view to limiting the width of cracks. By selecting an appropriate bearing system and appropriate reinforcement, the appearance of cracks wider than 0.1 mm should be reliably avoided.

● Connections to other parts of the construction, as well as joints resulting from interruptions in the work, should be avoided wherever possible. If this cannot be done, they must be designed and made water-tight (e.g. by inserting a suitable jointing strip).

● The water-proof concrete tanking construction must be continued to at least 500 mm above the highest expected water table. It is preferable to build up the water-proof concrete tanking to above the level of the site.

External basement wall
Sealing the external basement wall

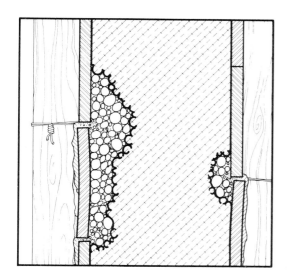

Damp damage to the external basement walls and water running into the basement rooms could in many cases be attributed to the porous structure of the concrete in external basement walls made of water-proof concrete. In places the concrete structure was very porous, there were pockets of gravel on the surface, there were timber spacers or reinforcing wires which penetrated the cross section and there were joints between individual sections of concrete that had not been specially protected.

Points for consideration

– In examples of a wall made of water-proof concrete, particularly high demands are placed on material selection and composition, on the mixing and application of the concrete and on its subsequent treatment.

– The type and composition of the additive material has a decisive influence on the water-proofness of the concrete structure. Porous, water absorbent additives and an irregular grain composition (sieving sequence) produce porous concrete because of the high porosity of the aggregate and gravel it contains. Therefore, the only materials which are suitable for use as additives are those with a dense and sufficiently solid structure and with a regular sieving sequence, and a grain size of up to 32 mm. In addition, a sealed structure and good workability depend on a sufficient proportion of powdered cement.

– The cement is the hydraulic binding agent for the concrete. The composition and strength of the cement is important for the strength of the finished concrete and the density of the concrete structure.

– When it is hydrated, cement binds with about 40 wt.% of water (chemically and physically). Thus when the ratio (by weight) of water to concrete is 0.4, the cement dust produces no pores in the concrete. However, as the proportion of water in the cement rises above 0.6 a system of pores is produced and this develops into a continuous capillary system. This makes the concrete highly porous. The proportion of water in the cement also affects workability (plasticity) and the compressibility of the concrete. By adding plasticisers, the proportion of water in the cement can be kept below 0.6.

– A concrete made up according to the above principles is generally already largely impermeable. Concrete sealing agents can further reduce water absorption by up to 20%. Here a distinction should be made between those agents which constrict pores by means of components which expand, those which interrupt pores by forming spherical pockets of air and those which render them water-proof by depositing hydrophobic agents.

– Any signs of separation in the concrete when it is applied, and all joints between adjacent parts of the structure or between adjacent materials represent non-water-proof areas. This also includes timber spacers and reinforcing wires, which remain in the cross section of the concrete, as well as insufficiently compressed concrete, pockets of gravel which are near to reinforcing rods and which do not allow the largest grains to pass through, and non water-proof walls, through which the cement lime can pass. In addition, the bond between concrete which has already set and freshly applied concrete is not homogeneous. If no precautions are taken, for example by inserting jointing strips or by using additives to retard setting, these joints in the work may lead to leaks.

External basement wall
Sealing the external basement wall

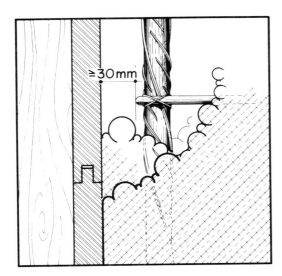

– Treatment of the concrete once it has been laid is particularly important. If the concrete dries out quickly as a result of excessively high air temperatures, or strong sunlight or wind, the quantity of water necessary for complete hydration is removed. This produces strength losses and a non-water-proof structure.

Recommendations for the avoidance of defects

● Water-proof concrete should be made from a dense aggregate with a maximum grain size of 16 or 32 mm with a graded sieving sequence, plus at least 350 kg/m³ of cement and an adequate proportion of fine grains. The proportion of water in the cement must be below 0.6. Concrete sealing agents can be used, provided that they have been proved to be suitable.

● The minimum cover of and the distance between the reinforcing rods, including the tie-bars, should be at least 30 mm and should be at least 5 mm greater than the maximum grain size of the additive.

● The concrete should be poured into the shuttering in such a way that separation cannot occur and a complete seal is achieved. No auxiliary materials that will remain in the concrete or that will be removed once the concrete has begun to set (racking wires) should be used.

● Work should be organised in such a way that there are no interruptions. Joints in the work should be avoided wherever possible, if necessary by adding retarding agents. Unavoidable joints in the work should be planned in terms of their position and construction, for example by incorporating a suitable joint gasket or strip, and by cleaning and preparing the surface of the concrete.

● Water-proof concrete structures should be protected from drying out and against frost. They should therefore be kept damp for at least twenty-one days and, if necessary, should be covered with thermal insulation mats. Water-proof concrete walls should be back-filled with soil as soon as possible after the shuttering has been removed. If this is not possible, the surface of the wall should be protected with several layers of a bituminous asphalt coating (see B 1.1.4).

● In view of the large number of factors that have to be taken into account in producing water-proof concrete, this work should only be carried out by specialist firms.

Problem: Structural cross section, thermal insulation layer, protective layer

Besides the work of damp protection, the external basement wall must perform other tasks in terms of the usefulness of the basement rooms and the effectiveness of the structural components.

The load of the building is transmitted to the foundations via the external basement walls, which are generally made of concrete, which can consist of reinforced or non-reinforced concrete. If the supporting cross section is unable to perform its supporting function without cracking this has a decisive influence on the effectiveness of the wall sealing, since the supporting wall in cross section is either responsible for the sealing function itself (water-proof concrete) or is the bearer and the support for the surface sealing layers.

Because of the increasingly demanding uses made of rooms with external walls that are in the ground, the thermal insulation function of the external basement wall becomes more and more important. In this respect, methods of providing adequate room and surface temperatures in order to reduce heating requirements and to prevent condensation on the wall surface become important.

The analysis of a representative sample of building damage showed that protective layers on top of surface sealing layers are necessary where walls are exposed to accumulated seepage water. Similarly, questions relating to the shape and the back-filling of the excavated site are also relevant, since they greatly affect the quality of the sealing work. The drainage layers in front of the external basement wall should also be considered in connection with this. This is dealt with separately in part A 1.3.

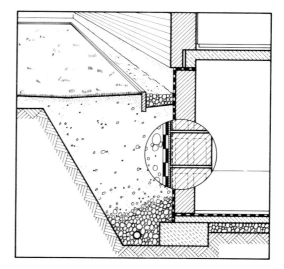

Because of the large number of types of damage which can occur to functional layers, either because of the layers themselves or because of their incorrect construction, and particularly damage to the sealing layers, it can be concluded that the attention paid to planning and execution is not commensurate with the heavy demands placed on the external basement wall.

The following pages show the recurring faults that occur in planning and execution and make recommendations for the avoidance of defects.

External basement wall

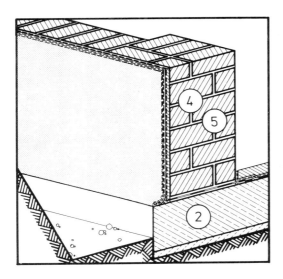

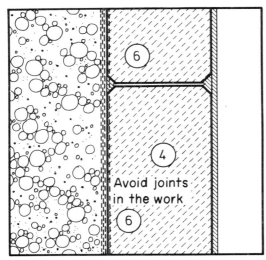

Avoid joints in the work

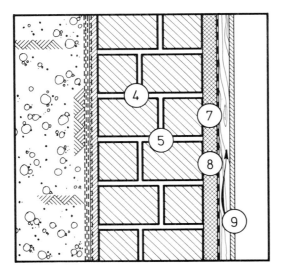

1 The size and foundations of the external basement walls should be such that the likelihood of wide cracks in the wall cross section caused by variations in settling is slight (see B 1.2.2).

2 In examples of long building structures, with alternating loads on the building or varying load carrying capacity of the soil, the building structure should either be built on a rigid slab of the correct size or should be divided by movement joints at the edges (or at maximum intervals of about 25 m (see B 1.2.2).

3 External basement walls that are in the ground must resist lateral pressure from the ground. The excavated site should not be back-filled until adequate strengthening has been provided to the wall by partitions, and until the basement ceiling slab has been constructed in such a way that it cannot be moved and until further loads (the subsequent storeys of the building) have been applied (see B 1.2.2).

4 The supporting wall cross section should be built as carefully as possible without leaks in order to minimise possible damage in the case of flaws in the external sealing measures (see B 1.2.3).

5 Brick walls should be made of bricks with a crushing strength of $\geq$ 5 MN/m² and mortars (MG II, IIa, III) whose consistency is as plastic as possible in order to prevent the appearance of cavities in the joints (see B 1.2.3).

6 Walls made of reinforced or non-reinforced concrete should be as dense as possible (proportion of water in the cement < 0.6, favourable grading sequence of the aggregate with sufficient proportion of fine grains) and sufficiently compressed and should be constructed in a rigid shuttering which is such that it will prevent irregularities on the surface of the concrete (see B 1.2.3).

7 The thermal transmission resistance of the external basement walls must be suitable for the purpose for which the basement rooms are going to be used. In the case of heated basement rooms, the thermal transmission resistance of walls which are completely below ground level should be at least 1.0m²K/W ($k \leq$ 0.90 W/m²K). With basement walls partially exposed to the open air, the thermal transmission resistance of the whole surface of the wall should be at least 1.3 m² K/W ($k \leq$ 0.68 W/m²K) (see B 1.2.4 and B 1.2.5).

8 The thermal insulation layer should be applied to the inside of the external basement walls. Wherever possible, the thermal insulation materials which are used should be waterproof and should have a high degree of resistance to water vapour (for example, expanded polystyrene sheets). Where the thermal insulation material is not vapour resistant (for example in the case of mineral fibre boards) a vapour barrier layer must be applied to the inside (for example laminated aluminium foil); building paper should then be used to seal the outside of the wall wherever possible (see B 1.2.5).

9 Claddings should be fixed to the internal surfaces of the external basement walls in such a way that efficient back ventilation is possible; the battens supporting the cladding should run vertically and there should be ventilation slits at least 20 mm wide at the base and at the top of the wall (see B 1.2.5).

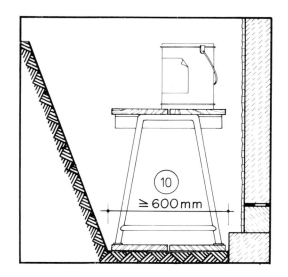

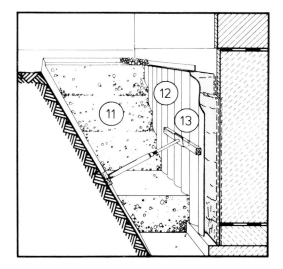

10 Before starting work on sealing external basement walls, the area excavated should be sufficiently wide (at least 600 mm at the bottom) and should be cleared away to approximately 100 mm below the top of the foundation slab. Where the soil is soft, two thick planks can be placed side by side and used as a working platform (see B 1.2.7).

11 The excavated site should be back-filled as soon as possible after the wall sealing measures have been completed and once any necessary drainage has been laid (see B 1.2.6 and B 1.2.7).

12 If sealed external walls have to be left for a relatively long time before back-filling, the wall surfaces should be protected against extreme climatic influences and excessive mechanical damage during construction work. Weather resistant timber planks or ply sheets which are sufficiently resistant to mechanical damage are suitable for this purpose (see B 1.2.6).

13 When the excavated site is back-filled, and particularly when this is done by machine, the sealing layer must be protected by boards or planks which are suitably resistant to mechanical damage (see B 1.2.6 and B 1.2.7).

14 A protective layer can only be dispensed with when the site is back-filled carefully with soil which has been selected bearing in mind factors connected with water conservation and drainage of the excavated site and which contains no stones or building debris, the soil being laid in layers 300 to 400 mm thick that is then compressed (see B 1.2.7).

External basement wall
Structural cross section, thermal insulation layer, protective layer

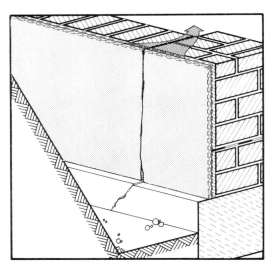

Many examples of damp damage could be attributed to flaws in mineral sealing layers or in water-proof concrete walls. These flaws in turn were the result of cracks or changes in the shape or length of the supporting wall in cross section.

Brick walls and walls made of non-reinforced concrete or of concrete which was inadequately reinforced were particularly affected by damage, as well as external walls consisting of sections with different loads or foundations. In some cases cracks were observed in the sealing layers, as well as horizontal displacement of sections of wall on a level with the horizontal damp-proof course in the external basement walls.

Points for consideration

– External basement walls made of brick and non-reinforced concrete, which are erected on non-reinforced strip foundations or on sectional raft foundation slabs, are particularly prone to cracking as a result of uneven building site loads, for example (i) between columns and infill wall panels as a result of soil with varying load-bearing strength, (ii) when the building ground varies between undisturbed soil and soil which has been back-filled, or (iii) because of varying degrees of lifting of the foundations caused by frost before the excavated site had been back-filled.

– External basement walls are exposed to lateral pressure from the ground. As the height of the back-filled wall increases, the pressure exerted by the ground also increases and this is particularly high in the case of non-cohesive soils, on slopes and where site traffic runs over the back-filled site. Failure to strengthen the external basement walls, by means of partition walls which tie in with them and by means of the ceiling slab and the load of further storeys, increases the risk of damage. Horizontal seals made of bituminous materials which are continuous through the wall's cross section may disturb the distribution of forces through the wall so that there is a danger that the wall sections will be displaced, particularly during the building period if the full weight of the building is not resting on the external basement wall.

– Brickwork that is mineral-based (e.g. sand-lime) and non-reinforced concrete walls are liable to undergo considerable stretching processes and are prone to inherent stresses as a result of shrinkage. In the case of relatively long walls, rapid drying and subsequent repeated damp penetration may result in cracks in the walls.

– Sealing layers made of rigid material applied to a substrate which is prone to cracking should be used only where they are exposed to soil moisture and non-accumulating seepage water because of the risk of possible damage. If a wall prone to cracking (maximum crack width 2 mm) is to be sealed against seepage water which accumulates for short periods, only seals which have sufficient stretching capacity and strength may be used.

Recommendations for the avoidance of defects

● The dimensions and foundations of external basement walls should be such that the likelihood of wide cracks appearing in the wall cross section as a result of irregular settling is slight.

● With long structural components, sudden variations in load or varying load-carrying capacity of the soil, the structural component should either be built on a rigid slab of the correct size or should be divided by settling joints at the edges.

● External basement walls in the ground must resist lateral pressure from the ground. The excavated site should not be back-filled until adequate strengthening has been provided by partition walls, until the basement ceiling slab has been constructed in such a way that it cannot be moved and until further loads have been applied.

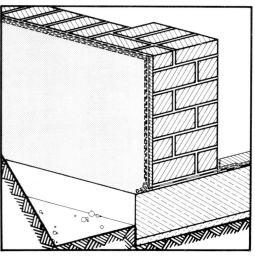

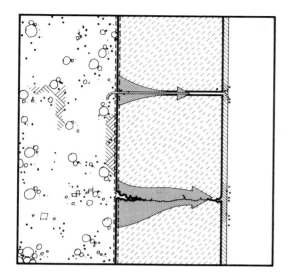

External basement wall
Structural cross section, thermal insulation layer, protective layer

If sealing layers began to leak because of flaws or because they were overloaded, the actual consequences of these leaks on the internal rooms were mainly dependent on flaws in the supporting wall cross sections. The points at which water appeared on the inside of the walls were near cracks and, in concrete walls, mainly near joints in the concrete, pockets of gravel, reinforcing wires and distance pieces, and in basement walls made of brick or blocks, particularly at their butt and bed joints.

In many examples, sealing layers failed because the substratum was defective. Irregularities in the wall surface to be sealed, flaws and open joints resulted in varying thicknesses of coating layers or coverings, or additional repairs and render layers were necessary, and these in turn were subject to problems (inadequate adhesion bond). Changes in material used in the wall surface, for example between supporting reinforced concrete columns and framed brickwork, also led to leaks or cracks.

Points for consideration

- In the presence of small quantities of water, the water penetrating the wall in cross section is absorbed and transported by capillary forces in the wall material. Larger quantities of water present in the soil may however penetrate the wall as a result of gravity or hydrostatic pressure, particularly when there are hollow cavities in the wall. In the event of exposure to seepage water, and particularly to accumulated water, heavy water penetration can be expected. This can also be expected where there are individual flaws in the sealing layer if the brick wall has bed and butt joints that are not completely filled so that the hollow cavities form an integrated system for penetration. This permeability to free water can be further increased by the holes or hollow cavities in the bricks.

- Whilst, for example, a typical wall brick has a large capillary absorptive capacity because of its pore structure, brickwork mortar from mortar group II, and especially group III, has a considerably lower capillary absorbency. Completely filled joints therefore inhibit capillary movement of water.

- Concrete in cross section with approved material composition and construction is impervious to free water. Hollow cavities or pore systems which penetrate the wall may be produced by reinforcing wires or distance pieces or at the joints between other parts of the structure or materials.

- The structural wall in cross section forms the substratum for the seal applied to the external basement wall. Seals applied in the form of surface coatings – bituminous coatings, water-proof concrete, sealing compounds – are particularly dependent on the nature of the wall surface, as the large numbers of cases of damage show. Whilst brickwork normally requires a smoothing coat, rendered layers themselves applied to concrete surfaces were often damaged and thus destroyed the seals.

- Seals consisting of bituminous coatings and mineral sealing materials are not completely water-proof. Thus, water can penetrate the cross section of the wall through flaws in the sealing layer. If the structural wall in cross section is not made of moisture resistant materials, this may result in serious damage to the supporting external basement walls which may under some circumstances have an adverse effect on the stability of the building. Parts of the structure that are outside the seal, such as the foundations, and are permanently damp, are particularly at risk. The base area of the external basement walls is exposed to the effects of frost in the external wall cross section. If this part of the structure

External basement wall
Structural cross section, thermal insulation layer, protective layer

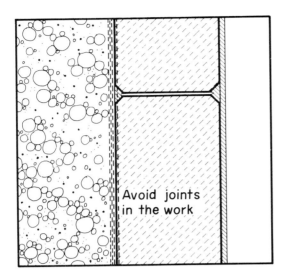

Avoid joints in the work

is made of a non-frost-resistant material which is damp, this may result in superficial structural damage. It is recommended that in the case of the external walls of the basement and of the base, brickwork made of bricks with a crushing strength of 5 MN/m² and mortars from groups II, IIa or III should be constructed to a height of 500 mm above the level of the site.

Recommendations for the avoidance of defects

● The structural wall in cross section should be constructed as carefully as possible without leaks in order to minimise possible damage in the case of flaws in the external sealing measures.

● Brick walls should be made of bricks with a crushing strength of > 5 MN/m² and mortars from mortar groups II, IIa, or III whose consistency is as plastic as possible in order to prevent the appearance of cavities in the joints.

● Walls made of reinforced or non-reinforced concrete should be as dense as possible (proportion of water in the cement < 0.6, favourable grading sequence of the aggregate with sufficient proportion of fine grains) and sufficiently compressed and should be constructed in a rigid shuttering which is such that it will prevent irregularities on the surface of the concrete.

External basement wall
Structural cross section, thermal insulation layer, protective layer

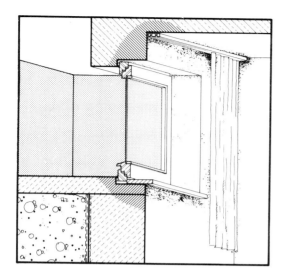

Damp internal wall surfaces, and black fungal growths resulting in damage to wallpaper and paintwork, particularly in the corners of walls, soffits and on parts of the surface which are not in contact with the ground, could be attributed to an inadequate thermal insulation value of the external basement walls. In some examples, the rooms concerned were heated, but had particularly high relative air humidity (for example bedrooms, bathrooms and drying rooms), but there were also living rooms only heated part of the time (for example, hobby rooms, play rooms and guest rooms), and in some cases unheated auxiliary rooms linked to the above mentioned rooms by doors which were permanently open.

Points for consideration

– Rooms with external basement walls which are below ground level and which are used by individuals as living rooms (hobby rooms, guest rooms, rooms used for commercial purposes) are particularly common in residential buildings built on slopes. This results in increased loads on parts of the construction as a result of moisture produced internally when the rooms are in use and in considerably heavier requirements being made on the parts of the construction in terms of thermal and moisture protection. A further factor to be considered is that of reducing heating costs, particularly in the case of rooms heated all the time.

– There is a particularly high risk of condensation forming on the inside of the wall if the external wall is made damp by water from the soil and if at the same time the relative air humidity inside the room is high. Sealing layers made of bituminous coatings, sealing compounds, water-proof render and water-proof concrete are not completely water-proof. The water which constantly penetrates the wall cross section through these sealing layers is carried to the inside of the wall under capillary action. Depending on the conditions (relative air humidity and air temperature) inside these rooms, this moisture reaches the internal surface of the wall (capillary action greater than drying capacity), or evaporates in the wall cross section (capillary action less than drying capacity).

– Because of their strength and stability, as well as their resistance to permanent exposure to water in the soil, concrete and building stone, which have good thermal conductivity, are used almost exclusively as materials for external basement walls. The thermal insulation value of such wall in cross section is, for example, approximately 0.65 m²K/W in the case of bricks (where the external basement walls are below ground level this is equivalent to a heat transfer coefficient of 1.3 m²K/W).

– As the moisture content of the wall cross section increases, the thermal insulation value falls. At points where there is little air movement, for example in corners, correspondingly low surface temperatures can be expected as a result of low thermal transmission resistance. Cold bridges caused by the construction or by the materials used (reinforced concrete lintels in a brick wall) also have low surface temperatures.

– If besides attempting to prevent surface water damage, efforts to produce a comfortable room climate and to reduce heat losses are taken into consideration, values considerably in excess of the minimum requirements are necessary. In the case of wall surfaces which are below ground level a thermal transmission coefficient of 0.90 W/m²K is recommended (this is equivalent to a minimum thermal resistance of 1.0 m²K/W). In the case of basement walls which are exposed to the outside air, the necessary thermal insulation

External basement wall
Structural cross section, thermal insulation layer, protective layer

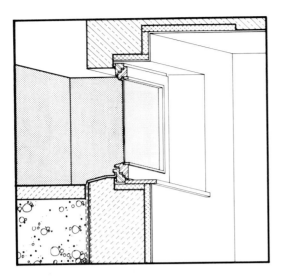

should be viewed in conjunction with the window surface areas. Economic considerations would indicate that values between 1.0 and 3.0 m²K/W are advisable.

– In examples of external basement walls where part of the wall is below ground level and part is above ground level there are considerable differences in terms of temperature loads. Although the calculated winter temperatures for external walls can be applied to the parts above ground level and the part of the wall down to a depth of about 800 mm, the remaining section of wall can be expected to have lowest winter temperatures in excess of 0°C.

Recommendations for the avoidance of defects

● The thermal transmission resistance of the external basement walls must be suitable for the purpose for which the basement rooms are going to be used. In the case of heated basement rooms, the thermal transmission resistance of walls completely below ground level should be at least 1.0 m²K/W ($k \leq$ 0.90 W/m²K). With basement walls partially exposed to the open air, the thermal transmission resistance of the whole surface of the wall should be at least 1.3 m²K/W ($k \leq$ 0.68 W/m²K).

External basement wall
Structural cross section, thermal insulation layer, protective layer

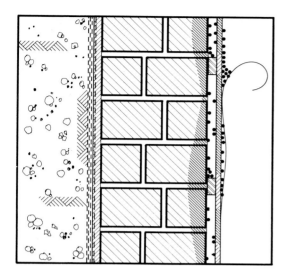

If the external walls of heated basements were covered with dry plaster or timber boards, the whole wall surface became discoloured and the wallpaper began to peel. Plaster board sheets and chipboard, as well as the supporting battens, were damp and covered with fungus in places. Painted wall surfaces were also damp. Similar faults appeared behind cupboards attached to external walls or behind curtains, etc. If light weight fibre boards were attached to the outside of the basement walls and then covered with plaster, this plaster and the sealing materials applied to it were in most cases largely torn and damaged, thus resulting in damp patches to the insides of the rooms.

Points for consideration

– Additional thermal insulation layers are necessary in order to produce a wall with an adequate thermal insulation value.

– In cases of thermal insulation layers applied to the insides of walls, harmful amounts of condensation may appear in the layer between the thermal insulation and the wall if the vapour barrier value (equivalent diffusion of a layer of air) is low and if the air humidity in the room is high. The quantity of water vapour that diffuses into the room can be reduced to a harmless level by using vapour-repellent layers or by selecting thermal insulation materials with a high water vapour diffusion resistance.

– External basement wall seals which involve bituminous coatings or water-proof render and sealing compounds are not completely water-proof. Similarly, water may penetrate the wall cross section through flaws in the sealing layer. Open-celled or fibrous thermal insulation materials, which have an additional vapour barrier, may become saturated from the outside in such cases. Only seals made of building paper or water-proof concrete are sufficiently water-proof.

– Layers of air trapped behind wall finishes or large units of furniture in the room act as thermal insulation layers. This effect is lessened as the movement of air in front of the surface of the wall increases.

– Thermal insulation materials normally have a low pressure resistance and this means that they may become seriously distorted by ground pressure and particularly by spot loads. Sealing layers which are applied to such thermal insulation materials are thus particularly at risk. Moreover, the sequence of layers used, which may be unfavourable where diffusion is concerned, may result in the harmful build up of condensation in the thermal insulation material.

– The position of the thermal insulation layer has a decisive effect on temperature conditions and on heating requirements in the case of basement rooms which are only used and heated for part of the time. Where the thermal insulation layer is arranged on the outside, the storage ability of the wall in cross section has an important effect on the temperatures inside the room. However, where the thermal insulation layers are arranged on the inside of the wall, the temperature stability of the rooms is reduced: they heat up and similarly cool down quickly.

– Thermal insulation materials which are incorporated into the outside of the external basement wall are constantly exposed to the effects of water in the soil. In particular, those thermal insulation layers of vegetable fibre and those with binding agents that swell or are not resistant to decay, rot away and become ineffective. Thermal insulation materials with open pores that absorb water under capillary action become saturated after a time and they thus lose most of their thermal insulation effect. Only fibre glass slabs and extruded

External basement wall
Structural cross section, thermal insulation layer, protective layer

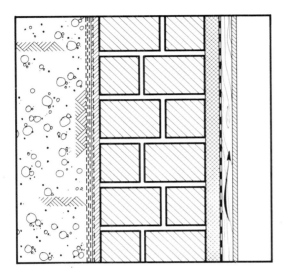

polystyrene foams do not seem to change their thermal conductivity when exposed constantly to water, because of their low absorbency. Because of insufficient long term experience with thermal insulation materials incorporated in this way layers of material arranged on the outside of sealing layers should not be included in calculating thermal transmission resistance.

Recommendations for the avoidance of defects

● The thermal transmission resistance of the external basement walls must be suitable for the purpose for which the basement rooms are going to be used. In the case of heated basement rooms, the thermal transmission resistance of walls completely below ground level should be at least 1.0 m²K/W ($k \leq 0.90$ W/m²K). With basement walls which are partially exposed to the open air, the thermal transmission resistance of the whole surface of the wall should be at least 1.3 m²K/W ($k \leq 0.68$ W/m²K).

● The thermal insulation layer should be applied to the inside of the external basement walls.

● Wherever possible, the thermal insulation materials which are used should be water-proof and should have a high degree of resistance to water vapour (for example, expanded polystyrene sheets). Where the thermal insulation material is not vapour resistant (for example, in the case of mineral fibre boards) a vapour barrier layer must be applied to the inside (for example, laminated aluminium foil); building paper should then be used to seal the outside of the wall wherever possible.

● Claddings should be fixed to the internal surfaces of the external basement walls in such a way that efficient back ventilation is possible; the battens supporting the cladding should run vertically and ventilation slits at least 20 mm wide should be arranged at the base and at the top of the wall.

External basement wall
Structural cross section, thermal insulation layer, protective layer

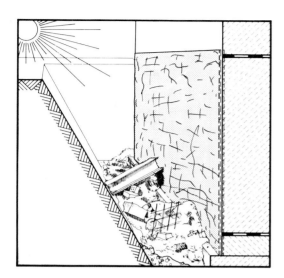

A large proportion of the damage to sealing layers could be attributed to the lack of protective measures in front of the external basement walls after they have been sealed. Both bituminous coatings and building papers, sealing compounds and water-proof renders were affected in this way. Protective layers or seepage layers, which at the same time fulfilled a protective function, were only used in a few examples. The following types of damage occurred: (i) Seals which had been left unprotected for a long time became damaged by impact loads as a result of building debris, etc. falling into the excavated site. (ii) Damage with subsequent damp penetration was particularly common where excavated sites were back-filled with a mixture of excavated soil and building debris, especially where this was done by machine. (iii) Adverse effects and damage were detected on the surface of basement walls that were exposed to climatic influences for a relatively long time. Fine cracks appeared in plasters and compounds, the surface of bituminous coatings became rigid, and building papers had pulled away from the top and had slipped, folding over in the process.

Points for consideration

– The surfaces of bituminous materials become weathered when constantly exposed to light and damp air alternately.

– When exposed to direct sunlight, black wall surfaces can reach very high temperatures (up to about + 70°C). When they reach such temperatures, bituminous coatings and especially the adhesive layers for bonded bituminous felts soften to such an extent that they begin to run and the layers begin to sag under their own weight unless this solar heat is reduced by shading, for example with protective boards or by white limewashes. Simultaneously such coverings applied to the black sealing material show up clearly any points where there have been mechanical pressures.

– Besides temperature fluctuations, moisture variations (rain/drought) also act on sealing layers in the form of stresses (temperature, shrinkage/expansion stresses) and these lead to cracks, especially in rendering and compound layers.

– If an excavated site is left for a relatively long time before being back-filled, unprotected wall surfaces may become damaged by falling objects, such as building debris, slabs, etc.

– Back-filling of the excavated site can be particularly harmful, especially in terms of damage to the sealing layers. When back-filling the site by machine, the impact load on the wall surface caused by the soil, which is sometimes tipped in large quantities and from a great height, is very great and it is highly likely to cause damage if the back-filling soil is mixed with stones or other sharp-edged objects. The mechanical pressure is reduced if non-cohesive soil is back-filled lengthwise, possibly instead of using a cohesive soil or excavated soil containing building debris.

– Protection against mechanical damage can be achieved by means of ply sheet or timber planks which are sufficiently rigid and resistant to chemical, physical and biological attack by the outside conditions and by the soil. Materials which need to be fixed to the wall surface by nails cannot be used. Protective layers which are adhered to the surface of the wall in strips or at individual points run the risk of peeling off if they are exposed to sunlight or water saturation.

External basement wall
Structural cross section, thermal insulation layer, protective layer

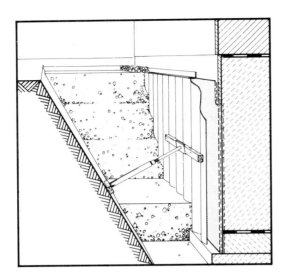

– In examples of protective layers which form hollow cavities, blockages might occur if soil or building debris fall inside them, and this in turn will mean that seepage water accumulates in front of the wall surface. This can be prevented if the upper edge is covered. Basically, a distinction should be made between protective layers and drainage or seepage layers in terms of their function. Seepage layers (e.g. drainage walls) act as a protective layer at the same time, whilst protective layers (e.g. corrugated sheets made of asbestos cement) are generally unable to perform any seepage function (see A 1.3).

Recommendations for the avoidance of defects

● The excavated site should be back-filled as soon as possible after the wall sealing measures have been completed.

● If sealed external basement walls have to be left a relatively long period before back-filling, the wall surfaces should be protected against extreme climatic influences and excessive mechanical damage during construction work. Weather resistant timber planks or ply sheets that are sufficiently resistant to mechanical damage are suitable for this purpose.

● When the excavated site is back-filled, and particularly when this is done by machine, the sealing layer must be protected by boards or planks suitably resistant to mechanical damage.

External basement wall
Structural cross section, thermal insulation layer, protective layer

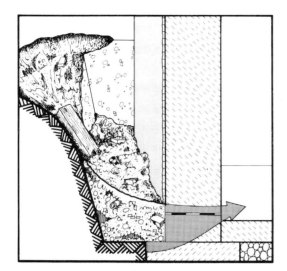

Incorrectly constructed external sealing layers applied to external basement walls could be partially attributed to bad site organisation. This also included cases where the external rendering had not been applied correctly when constructing the wall from the inside because the excavated site was too narrow, or where there was no seal at the bottom of the wall, because when the sealing work was carried out the excavated area was partly filled with soil and debris.

In a large number of examples, sealing layers applied to the external basement walls became damaged by impact loads during the building work or during back-filling of the excavated site. This included both damage caused by building debris and rubble which fell into the excavated site before back-filling and damage caused by back-filling itself, where material containing stone and building debris was back-filled mechanically in one operation.

Points for consideration

– Because of the fundamental need for a dry, firm substratum without changes in shape or length which might bring about cracking, external sealing layers should be applied as late as possible after the basement roof slab has been fitted and after most of the upper storeys have been completed.

– The composition of the surrounding soil will allow a specific maximum angle of repose for the soil. If the wall of the excavation is too steep, it will cave in, particularly if it rains.

– Heavy rain implies that large quantities of water will run along the surface into the excavation particularly where the site is on a slope. This water will carry with it fine soil particles, which are deposited on the floor of the excavation where they form a soil layer with low permeability. If this soil is not removed before the site is back-filled, damp can be expected to accumulate in front of the basement wall, even where the rest of the surrounding soil is permeable.

– During building, it is virtually impossible to prevent building debris and other heavy objects from falling into the excavated site, some of them from a great height, and thereby striking the basement wall. Externally applied surface sealing layers can be damaged in this way.

– In order to ensure that the sealing work is carried out correctly, a minimum amount of working space is necessary, as well as a firm and clean base and the ability to erect scaffolding. Before starting sealing work, it is therefore normally necessary to prepare the excavated site accordingly.

– Back-filling of the excavated site becomes particularly important in terms of the water conditions in front of the external basement wall and of the effectiveness of the drainage measures which might be used (see A 1.3). However, in order to avoid overloading the sealing layers, the back-filling material, the type of back-filling used and any measures for protecting the wall surface must be chosen carefully. The impact load of the large quantity of back-filled soil, possibly filled from a great height, is particularly high in mechanical back-filling. There is a danger that a sealing layer will become damaged in this way as a result of abrasion, and particularly as a result of impact loads caused by sharp edged objects (stones, building debris). When using non-cohesive material, (e.g. gravel), and particularly if the material is filled in layers (300-400 mm thick) and compressed, the mechanical load can be controlled and is considerably less.

External basement wall
Structural cross section, thermal insulation layer, protective layer

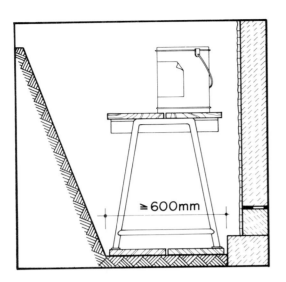

Recommendations for the avoidance of defects

● Before starting work on sealing external basement walls, the excavated site should be sufficiently wide (at least 600 mm at the bottom) and should be cleared away to approximately 100 mm below the top of the foundations. Where the soil is soft, two thick planks can be placed side by side and used as a working platform.

● The excavated site should be back-filled as soon as possible after the wall sealing measures have been completed and once any necessary drainage has been laid.

● When the excavated site is back-filled, and particularly when this is done by machine, the sealing layer must be protected by boards or planks which are suitably resistant to mechanical damage.

● A protective layer can only be dispensed with when the site is back-filled carefully with soil containing no stone or building debris in layers 300 to 400 mm thick, each layer being compressed.

● The back-filling soil should be selected bearing in mind factors connected with water conservation and the drainage of the excavated site (see A 1.3).

General specialist literature

Albrecht, Rudolf: Bauschäden – Vermeiden, Untersuchen, Sanieren. Bauverlag, Wiesbaden und Berlin 1976.

Lufsky, Karl: Bauwerksabdichtung, Bitumen und Kunststoffe in der Abdichtungstechnik. 3. Auflage. B. G. Teubner, Stuttgart 1975.

Reichert, Hubert: Abdichtungsmaßnahmen an erdberührten Bauteilen im Wohnungsbau. Forum-Fortbildung Bau, Forum-Verlag, Stuttgart 1977, Heft 8, Seite 101–114.

Reichert, Hubert: Sperrschicht und Dichtschicht im Hochbau. Verlagsgesellschaft Rudolf Müller, Köln 1974.

Schild, Erich: Untersuchung der Bauschäden an Kellern, Dränagen und Gründungen. Forum-Fortbildung Bau, Forum-Verlag, Stuttgart 1977, Heft 8, Seite 49–67.

AIB: Anweisung für die Abdichtung von Ingenieurbauwerken. Deutsche Bundesbahn, 2. Ausgabe 1953.

Sealing the bituminous coatings

Bierhalter, W.: Bitumenemulsionen für den Bautenschutz. In: Das Baugewerbe, Heft 17/1973, Seite 42–46.

Depke, Fritz, M.: Sicherer Schutz von Keller-Außenwänden. In: Das Baugewerbe, Heft 10/1976, Seite 24–26.

Fuhrmann, Wilfried (Hrsgb).: Bitumen- und Asphalt-Taschenbuch. 5. Auflage, Bauverlag, Wiesbaden und Berlin 1976.

Gundermann, Erich: Bautenschutz, Chemie und Technologie. 2. Auflage, Verlag Theodor Steinkopf, Dresden 1970.

Hegemann, Franz: Bitumenemulsionen und -lacke für den Bautenschutz. In: Baugewerbe, Heft 13/1976, Seite 22–24.

Moritz, Karl: Bauwerkssperrschichten und -abdichtungen im Hochbau. In: Deutsche Bauzeitschrift (DBZ), Heft 7/1963, Seite 1029–1032.

Setzer, Lothar; Richter, Heinz: Erdberührte Bauteile, Abdichtung mit bituminösen Stoffen gegen nichtdrückendes Wasser. In: Das Baugewerbe, Heft 10/1976, Seite 37–38.

DIN 4117: Abdichtung von Bauwerken gegen Bodenfeuchtigkeit, Richtlinien für die Ausführung. November 1960.

DIN 4122: Abdichtung von Bauwerken gegen nichtdrückendes Oberflächenwasser und Sickerwasser mit bituminösen Stoffen, Metallbändern und Kunststoffolien, Richtlinien. Juli 1968.

DIN 18195 – Teil 2 (Entwurf): Bauwerksabdichtungen, Stoffe. November 1977

DIN 18195 – Teil 4 (Entwurf): Bauwerksabdichtungen, Abdichtung gegen Bodenfeuchtigkeit, Ausführung und Bemessung. November 1977

DIN 18337 – VOB, Teil C: Abdichtung gegen nichtdrückendes Wasser. Februar 1961.

Sealing with water-proof rendering

Albrecht, Walter; Mannherz, Ursula: Zusatzmittel, Anstrichstoffe, Hilfsmittel für Beton und Mörtel. 8. Auflage, Bauverlag, Wiesbaden und Berlin 1968.

Piepenburg, Werner: Mörtel, Mauerwerk, Putz. 6. Auflage, Bauverlag, Wiesbaden und Berlin 1970.

DIN 4117: Abdichtung von Bauwerken gegen Bodenfeuchtigkeit, Richtlinien für die Ausführung. November 1960.

DIN 18550: Putz, Baustoffe und Ausführung. Juni 1967.

Sealing with sealing compounds

Brand, Hermann: Die zementgebundenen Oberflächendichtungsmittel. In: Das Baugewerbe, Heft 10/1976, Seite 35–36.

Grunau, Edvard B.: Bauwerksabdichtung – Möglichkeiten und Methoden. In: Baugewerbe, Heft 18/1976, Seite 26–36.

Köneke, Rolf: Bauwerksabdichtungen nach DIN 4117 und DIN 18550 oder mit Zementschlämmen oder womit? In: Das Baugewerbe, Heft 20/1975, Seite 36–40.

Köster, Johann J.: Flächenabdichtungen bei Bauwerken. In: Das Baugewerbe, Heft 9/1975, Seite 31–32.

Schumann, Dieter: Untersuchung über die Wirksamkeit von Dichtungsschlämmen; Abschlußbericht zu einem Forschungsauftrag des Bundesministeriums für Raumordnung, Bauwesen und Städtebau, Lehrstuhl für Baustoffkunde und Werkstoffprüfung der Technischen Universität München. München, April 1977.

Schumann, Dieter: Sperr- und Sanierungsputze als flankierende Maßnahmen bei der Mauerwerksentfeuchtung. Vervielfältigtes Vortragsmanuskript, Mai 1977, Bauzentrum München.

Verband der Bautenschutzmittel-Industrie: VdB Standard Dichtungsschlämmen, Technisches Merkblatt zur Abdichtung von Bauwerken gegen drückendes und nichtdrückendes Wasser durch werksmäßig vorgefestigte, zementgebundene, dünnschichtige Oberflächendichtungsmittel. Frankfurt.

Internal sealing

Brand, Hermann: Nachträgliche Abdichtung auf chemischem Wege. Forum-Fortbildung Bau, Heft 8, Forum-Verlag, Stuttgart 1977, Heft 8, Seite 86–88.

Horstschäfer, Heinz-Josef: Nachträgliche Abdichtungen mit starren Innendichtungen, Forum-Fortbildung Bau, Heft 8, Forum-Verlag, Stuttgart 1977, Heft 8, Seite 82–85.

Schild, Erich: Nachbesserungsmaßnahmen bei Feuchtigkeitsschäden an Bauteilen im Erdreich. Forum-Fortbildung Bau, Heft 8, Forum-Verlag, Stuttgart 1977, Heft 8, Seite 76–81.

Sealing with building papers

Hauptverband der Deutschen Bauindustrie, Bundesfachabteilung Bauwerksabdichtung: Technische Regeln für die Planung und Ausführung von dehnfähigen Bauwerksabdichtungen. Otto Elsner-Verlagsgesellschaft, Darmstadt 1974.

Braun, E., Metelmann, P.; Thun, D.: Bituminöse Hautdichtungen – Folgerungen aus Theorie und Praxis. In: Bitumen, Heft 5/1973, Seite 117–129.

DIN 4031: Wasserdruckhaltende bituminöse Abdichtungen für Bauwerke, Richtlinien für Bemessung und Ausführung. November 1959.

DIN 18195 – Teil 6 (Entwurf): Bauwerksabdichtungen, Abdichtungen gegen von außen drückendes Wasser, Ausführung und Bemessung. November 1977

DIN 4122: Abdichtung von Bauwerken gegen nichtdrückendes Oberflächenwasser und Sickerwasser mit bituminösen Stoffen, Metallbändern und Kunststoffolien, Richtlinien. Juli 1968.

DIN 18336 – VOB, Teil C: Abdichtungen gegen drückendes Wasser. Oktober 1965.

DIN 18337 – VOB, Teil C: Abdichtung gegen nichtdrückendes Wasser. Februar 1961.

Sealing with water-proof concrete

Albrecht, Walter; Mannherz, Ursula: Zusatzmittel, Anstrichstoffe, Hilfsstoffe für Beton und Mörtel. 8. Auflage, Bauverlag, Wiesbaden und Berlin 1968.

Albrecht, Walter: Über die Wirkung von Betondichtungsmitteln. In: Betonstein-Zeitung, Heft 10/1966, Seite 568–573.

Breuckmann, K.: Herstellen von Beton mit Betonzusatzmitteln. In: Baugewerbe, Heft 15/1976, Seite 14–17.

Gunau, Günter; Klawa, Norbert: Empfehlungen zur Fugengestaltung im unterirdischen Bauen. In: Die Bautechnik, Heft 10/1973, Seite 325–332.

Karsten, Rudolf: Der heutige Stand der Anwendung von dichtenden Zusätzen zu Mörtel und Beton. In: Das Baugewerbe, Heft 9/1971, Seite 782–786.

Leonhardt, Fritz: Über die Kunst des Bewehrens von Stahlbetontragwerken. In: Beton- und Stahlbetonbau, Heft 8/1965, Seite 181–192.

Rapp, Günter: Technik des Sichtbetons. Beton-Verlag, Düsseldorf 1969.

Wesche, Karlheinz: Baustoffe für tragende Bauteile, Band 2, Nichtmetallische – anorganische Stoffe, Beton und Mauerwerk. Bauverlag, Wiesbaden und Berlin 1974.

Wischers, G.: Zur Wirksamkeit von Betondichtungsmitteln. In: beton, Heft 8/1975.

DIN 1045: Beton- und Stahlbetonbau, Bemessung und Ausführung. Januar 1972.

Structural cross section

DIN 1045: Beton- und Stahlbetonbau, Bemessung und Ausführung. Januar 1972.

DIN 1053 – Blatt 1: Mauerwerk, Berechnung und Ausführung. November 1974.

DIN 1054: Baugrund, Zulässige Belastung des Baugrundes. November 1969.

DIN 1055 – Blatt 2: Lastannahmen für Bauten, Bodenwerte, Berechnungsgewicht, Winkel der inneren Reibung, Kohäsion. Juni 1963.

DIN 4019 – Blatt 1: Baugrund, Setzungsberechnungen. September 1974.

External basement wall
Typical cross section

Thermal insulation

DIN 4108: Wärmeschutz im Hochbau. August 1969. Ergänzende Bestimmungen. Oktober 1974.
Wärmeschutzverordnung: Bundesgesetzblatt 1977, Teil I. Seite 1554ff.
Wallmeier, Jörg-Rüdiger: Nutzungsmöglichkeiten von Räumen mit erdberührten Außenwänden. Dissertation an der Fakultät für Bauwesen der RWTH Aachen, 1974.

Protective layer, building pit

Muth, Wilfried: Dränung zum Schutz von Bauteilen im Erdreich. Forum-Fortbildung Bau, Forum-Verlag, Stuttgart 1977, Heft 8, Seite 115–127.
Muth, Wilfried: Schutzschichten vor erdberührten Wänden. In: Deutsche Bauzeitschrift (DBZ), Heft 8/1976, Seite 1025–1027.
Rogier, Dietmar: Schäden und Mängel am Dränagesystem. Forum-Fortbildung Bau, Forum-Verlag, Stuttgart 1977, Heft 8, Seite 68–75.

Problem: Connection to the basement floor

When considering protection against the action of water from the soil, the external basement wall and the basement floor must be thought as one together with the connecting structures which join them. Detailed constructions are needed in connecting the base area with openings, light shafts, external basement stairs and apertures, and to the basement floor.

Solving the problem of connecting the vertical sealing layers of the external basement wall to those of the horizontal basement floor becomes particularly significant. The heaviest loads occur at the base of the external wall, and in conventional types of building it is not possible to make a direct connection between the sealing layers applied to the outside of the basement wall and to the floor slab because the external basement wall is constructed on the foundation or on the floor slab and the sealing layer would have to be continuous through the wall.

A direct connection between the wall and floor seals can be dispensed with if the basement rooms are for secondary use and if they are exposed only to soil moisture. In this case a horizontal damp-proof course can be used to give protection against moisture rising through the wall in cross section under capillary action. However, if the wall is exposed to water under pressure, particular care and effort will have to be devoted to designing and executing this point of detail in connection with the 'tanking', because of the serious consequences of damage. However, in the majority of cases where more demanding uses are made of basement rooms and where the excavated site is exposed to accumulated water, too little attention is paid to the construction of this detail. Inadequate methods of construction can thus be expected.

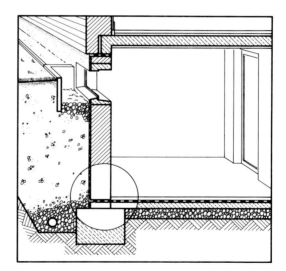

The frequency and extent of the damage in the connection between the external basement wall and the basement floor is correspondingly high. Since the lower horizontal damp-proof course is basically a continuation of the floor seal through the wall in cross section, and because the correct connection of these parts has a particularly important influence on the arrangement, material selection and construction of the wall seal, the recurring faults observed in the construction of this point of detail are dealt with and analysed in the appendix to section C 2.1, and means of avoiding them are recommended.

Problem: Structural wall and basement ceiling slab abutment

The base area forms the transition between the part of the basement wall which is in the ground and the external wall above ground level. Although in most cases this area is not visible, this part of the external basement wall must be built according to certain building principles in order to ensure that the construction will be permanently free from damage. It differs, in terms of its construction and of the materials used, both from external walls above ground and from those below ground level. This is justified by the special pressures to which this section of wall is subjected. In addition to the pressures on the external walls above ground caused by extreme temperatures, heavy rain, sunlight, etc., and in addition to the constant exposure to water which walls below ground level suffer, this particular section of the wall is subjected to additional pressures from spray water, puddles, impact loads and chemical attack. In addition, the constructional form of the basement ceiling slab abutment and its height affect the base construction.

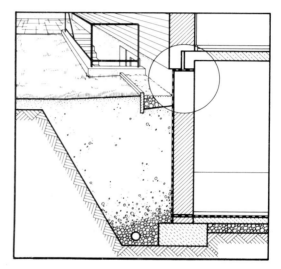

The damage observed in base areas reflects the transitional function of this section of the wall. In addition to damage resulting from material selection and from the construction of the base area itself, damage could also be attributed to incorrectly built supporting base for the rising external walls, as described in Volume 2 of this series, and to faults in the sealing of external basement walls below ground level. Similarly, the form and levels of the site surface were also reasons for damage.

These individual groups of problems are listed below, analysed and recommendations are made for avoiding damage.

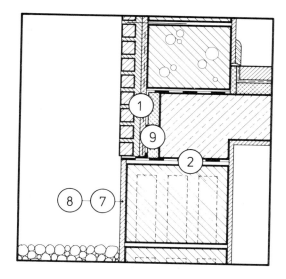

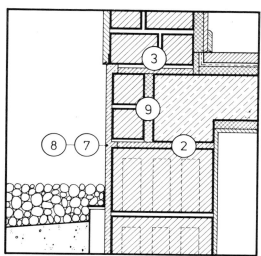

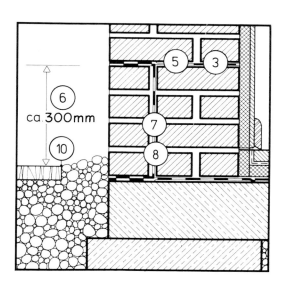

1 External walls above ground with claddings applied with mortar and with facing leaves must have a horizontal seal, for example made of one layer of building paper or bituminous felt, at the point where they meet with the basement ceiling slab or external basement wall. This seal must be placed from the outside of the facing leaves or cladding, from which point it should be turned back by at least 100 mm against the inner wall leaf on a level with the wall and slab joint and should pass through to the inside of the wall. In cases of two leafed external walls with a cavity, the horizontal seal in the inner wall leaf can be dispensed with (see B 2.2.2).

2 A horizontal seal (damp-proof course) must be installed below the basement ceiling slab continuously through the whole of the wall in cross section including any surface layers. This seal should be connected to the vertical wall or base seal in such a way that moisture bridges are avoided (see B 2.2.3).

3 If the horizontal wall seal (damp-proof course) below the basement ceiling slab is at a height where it might be affected by spray water, a further horizontal seal must be fitted in the wall above the height of the spray water (see B 2.2.3).

4 In the case of two leaf brickwork with a cavity, whose facing leaf continues as far as the spray water area, a further horizontal seal must be fitted in the external leaf above the spray water height (see B 2.2.3).

5 As a rule, the upper horizontal wall seal (damp-proof course) should be made of building paper, bituminous felt or plastic film, which should be fitted loosely with at least 100 mm overlap at the joints. The contact surface of the sealing layer should be levelled with a mortar smoothing coat (MG III) (see B 2.2.3).

6 The top of the base area is limited by the upper horizontal wall seal. The minimum height of the base area is equivalent to the spray water height; this should be 300 mm. However, where the surface of the site is harder and level, it should be raised to approximately 500 mm, whilst if a 500 mm wide strip of gravel is provided and if the site slopes away from the base, it can be lowered to about 200 mm (see B 2.2.4).

7 The base area of the external wall must have a vertical seal on the surface or in the wall cross section and this seal must connect with the vertical seal of the underground external basement walls and to the horizontal wall seals in such a way that moisture bridges are avoided (see B 2.2.4).

8 Vertical seals in the base area can be made of reinforced water-proof concrete at least 100 mm thick or water-proof rendering and of sealing courses or plastic films, which should be protected by a facing wall at least half a brick thick (for example of whole sand-lime bricks). Sealing plasters, several layers of bituminous coatings or water-proof applied thin coatings, as well as brickwork made of clinker bricks and cement mortar (MG III) or small tiles which have been applied with mortar should not be used as the sole form of sealing (see B 2.2.4).

9 Materials that are not weather or moisture resistant (e.g. thermal insulation layers) should be applied to the inside of the vertical sealing layer. Only frost-resistant materials should be applied to the outside (see B 2.2.4).

External basement wall

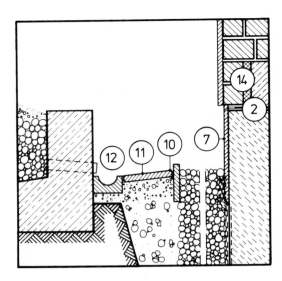

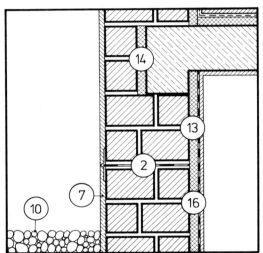

10 The height of the final site surface along the external walls of the building must already be fixed at the design stage (see B 2.2.5).

11 The surface of the excavated site should where possible slope away from the external wall by about 5% once it has been back-filled in layers and compressed. If a fairly large amount of settling is to be expected, for example where wet cohesive soil is back-filled, the site should be over-filled accordingly (see B 2.2.5).

12 Water being fed from the surrounding site to the external walls of the building should be avoided wherever possible, for example, by trapping the water early in open drainage channels parallel to the building (see B 2.2.5).

13 The thermal transmission resistance of the external basement walls must be suitable for the intended use of the basement rooms. In the case of heated basement rooms, part of which are in contact with the open air, the whole surface should have a thermal transmission resistance of at least $1.3 \ m^2K/W$ ($k \leq 0.68 \ W/m^2K$) (see B 2.2.6).

14 The external faces of the basement ceiling slabs near the abutment with the external basement walls should be protected by thermal insulation layers with a thermal transmission resistance of at least $1.3 \ m^2K/W$ ($k \leq 0.68 \ W/m^2K$).

15 If it is not possible to apply external thermal insulation to the external face of the basement ceiling slab, 500 mm wide strips of thermal insulation material with a thermal transmission resistance of at least $0.65 \ m^2K/W$ ($k \leq 1.24 \ W/m^2K$) should be applied to the underside of the ceiling slabs along the external walls (see B 2.2.6).

16 In the base area, the thermal insulation layers should be applied to the external basement walls or should be fitted in the wall cross section on the inside of the internal seal. The thermal insulation materials used should be moisture resistant and should have a high resistance to water vapour diffusion. Where the thermal insulation material will not prevent the passage of water vapour, a vapour barrier must be applied to the inside (see B 2.2.6).

External basement wall
Structural wall and basement ceiling slab abutment

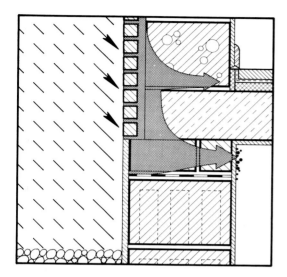

Some of the damp damage to the inside of buildings in the upper area of the external basement wall, as well as damp damage to the base of the external walls of the ground floor, with related signs of efflorescence on the outside, could not be attributed to the effect of damp in the soil or to spray water (see B 2.2.3). In these cases rainwater was the direct cause of the damage, particularly where heavy rain fell against walls with claddings and facing leaves, and in some cases against walls made of thin masonry slabs. Similar signs of damp penetration were found in projecting rendered base areas, but in these cases the damp penetration was linked with damage to the base render.

Points for consideration

– A film of water forms over the façade of the building during heavy rainfall. If this trickle of water is allowed to accumulate, for example by a projecting support, and if this support is not built to withstand this pressure (for example with metal flashings or angled bricks behind the façade cladding) a relatively large amount of water will penetrate the wall cross section at this point. This will result in fairly prolonged dampness, efflorescence and in expansion and shrinkage, as well as in frost action causing cracks and chips in the render.

– Brickwork, facing leaves and small unit claddings applied with mortar are not completely water-proof even if the bricks and mortar used are water-proof. Rainwater, particularly in heavy rain, penetrates the wall cross section under capillary action through joints or through the bricks and through unavoidable cracks in the joints. In the case of larger cracks or hollow joints it can also flow or seep directly into the wall cross section.

– The water which has penetrated the wall travels downwards under the force of gravity, this process being promoted by the continuous, channel-type cavities which can form in air layers, inadequately filled brick joints, thermal insulation layers or in perforations in the bricks. At the point where the external wall or the facing leaf make contact with the basement ceiling slab or other parts of the structure which will arrest seepage, the water will accumulate and will penetrate the inside or outside of the structure, resulting in damp in the render or the floor construction, or in deposits of substances carried in the water.

– Further details on this problem can be seen in Volume 2 of this series.

Recommendations for the avoidance of defects

● External walls above ground with claddings applied with mortar and with facing leaves must have a horizontal seal, for example made of a layer of building paper or bituminous felt at the point where they meet with the basement ceiling slab or external basement wall.

● The horizontal base seal can, for example, be made of a sealing course which is placed from the outside of the facing leaf or cladding towards the inside, from which point it should be turned back by at least 100 mm against the inner wall leaf on a level with the wall and slab joint and should pass through to the inside of the wall.

● In the case of two leaf external walls with a cavity, the horizontal seal in the inner wall leaf can be dispensed with.

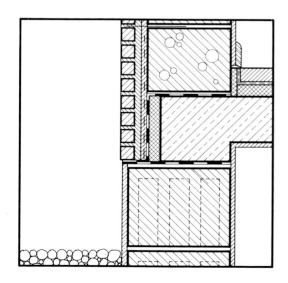

External basement wall
Structural wall and basement ceiling slab abutment

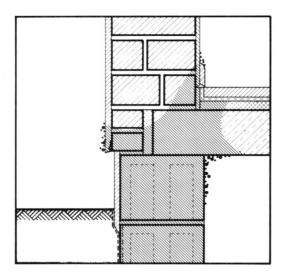

Some of the damp damage to the base area could be attributed to the effect of soil moisture or spray water. This affected the upper area of the external basement walls and sometimes also the base of the ground floor external walls. On the outside of the brickwork or render, strips along the base remained damp for prolonged periods. The top edges of these strips were irregular and showed signs of efflorescence and damage to the render (see B 2.2.2). In these cases there was either a complete lack of horizontal seals in the wall in the area of the basement ceiling slab, or they were only marginally above the site surface or were in the ground. In addition, external walls whose ventilated facing leaves were continued as high as the surface of the site were also affected.

Points for consideration

– Water which has penetrated a wall's cross section spreads out under capillary action. Besides the harmful effects of damp penetrating the inside of the wall, it is particularly damaging if the water is able to penetrate the basement ceiling or the floor or walls of the ground floor. The quantity of water carried and the height to which it rises depends mainly on the quantity of water which is absorbed and also on the pore structure, i.e., the capillary conductivity of the wall materials (bricks, mortar, render). Materials which have a large pore volume in the form of continuous pore systems (bricks, sand-lime bricks, lime cement mortar) are highly conductive from a capillary point of view. Materials with a low pore volume or with predominantly spherical and largely closed individual pores (cement mortar, high grade concrete or aerated concrete) are therefore less conductive from a capillary point of view.

– The vertical surface sealing layers of the external basement walls or of the base area can be expected to have some flaws as a result of incorrect construction or damage. During construction as well as during the service life of the building, water from the soil or spray water can be expected to be absorbed through these flaws into the wall cross section until the failures are repaired. In addition, the water pressure may increase during the service life of the building.

– Cross section seals incorporated into the basement walls, which continue through the whole of the wall cross section (including the surface layers) and which are connected to the vertical surface seal in such a way that moisture bridges are prevented, can limit this type of damp damage and can protect structural components which are sensitive to damp. Materials which will not transport water under capillary action and sealing layers are suitable for this purpose.

– Work involving the subsequent incorporation of a horizontal seal into the wall cross section is very costly and is not always possible. For this reason, the materials used for making this seal (permanent, insensitive to water) and the way in which it is made (single layer, resistant to mechanical damage) should be such that the risk of damage is kept as low as possible. Seals made of building paper, bituminous felt or plastic films with carefully placed overlapped joints are easier to check for effectiveness during building work than seals made of water-proof mortar or sealing compounds.

External basement wall
Structural wall and basement ceiling slab abutment

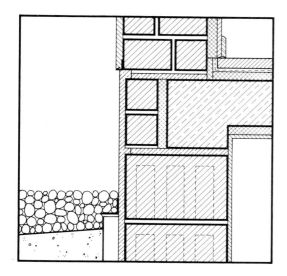

Recommendations for the avoidance of defects

● The height of the final site surface along the external walls of the building must already be fixed at the design stage.

● A horizontal seal (damp-proof course) must be installed below the basement ceiling slab continuously through the whole of the wall in cross section including any surface layers. This seal should be connected to the vertical wall or base seal in such a way that moisture bridges are avoided.

● If the horizontal wall seal (damp-proof course) below the basement ceiling slab is at a height where it might be affected by spray water (see B 2.2.4), a further horizontal seal must be fitted in the wall above the height of the spray water.

● In the case of two leaf brickwork with a cavity whose facing leaf continues as far as the spray water area, a further horizontal seal must be fitted in the external leaf above the spray water height.

● As a rule, the upper horizontal wall seal (damp-proof course) should be made of building paper, bituminous felt or plastic film, which should be fitted loosely with at least 100 mm overlap at the joints. The contact surface of the sealing layer should be levelled with a mortar smoothing coat (MG III).

External basement wall
Structural wall and basement ceiling slab abutment

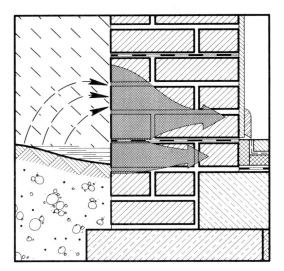

Buildings which have no wall seals or inadequate wall seals against exposure to water from the outside in the base area showed signs of damage, both on the inside and on the outside.

Inside, the damage appeared in the ground floor walls below, and depending on the level of the site surface above, the basement ceiling. Dirty areas appeared outside and where façade coatings were continued as far as the base area there were cracks, bubbles and peeling. Where external wall render and small tile claddings in mortar were applied there were also cracks and peeling over large areas, whilst in examples with thin masonry slabs and facing bricks there were signs of efflorescence.

Points for consideration

– On a level with the surface of the site, the external wall is exposed to particular pressures which differ from those of the external wall above ground and of the basement wall which is below ground level. These pressures require special consideration to be applied to construction and to the selection of material in the base area.

Besides the soil moisture which has an effect in the ground, corrosive substances (humic acid, thawing salts) may also be present in the water. If the shape of the site is unfavourable (slope down towards the wall) the flow of water towards the wall may be increased and this will lead to the formation of puddles directly in front of the outside wall. Spray water up to a height of about 500 mm can be expected, particularly when the site consists of a hard, level surface (slabs). Where the surface is uneven and likely to promote good dispersion (loose gravel filling diameter 32–63 mm), the spray water height is reduced to about 200 mm. Where the building has an open aspect, snow drifts will be formed against the wall of the building and this will result in melting water pressure at a corresponding height. If traffic surfaces (pathways, court-yards) were adjacent to the building, this resulted in increased mechanical damage under some circumstances (impact).

– The height of the base area to be protected is normally deter-mined by the height of the spray water pressure and extends as far as the upper horizontal seal of the external wall (see B 2.2.3). Because of the above-mentioned pressures, only materials which are permanently moisture and frost resistant, as well as being sufficiently strong and resistant to mechanical damage, are suitable for the base area. A vertical seal must be provided in order to protect components and materials that are moisture sensitive.

– Where sufficiently resistant surface seals (water-proof render) or water-proof concrete are used the external basement wall construction can be continued up as far as the upper horizontal cross section seal. Other types of seal (sealing courses, sealing compounds) are too sensitive and may not be continued up above site level unless they are protected.

– For aesthetic reasons, a certain degree of building up of the surface of the base area may be desirable, for example in brickwork. If this is done a connection is necessary between the sealing layers where the basement wall joins the base area in order to prevent moisture bridges. The water running down the façade surface may run in behind the seal, particularly in the case of sealing courses if the upper edge abuts the wall surface.

– In order to prevent moisture bridges, the vertical wall or base seal must be joined to the horizontal wall seals.

External basement wall
Structural wall and basement ceiling slab abutment

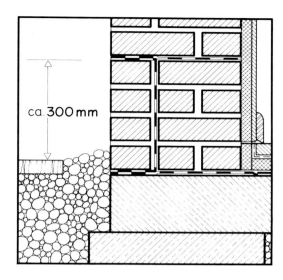

Recommendations for the avoidance of defects

● The height of the final site surface along the external walls of the building must already be fixed at the design stage.

● The top of the base area is limited by the upper horizontal wall seal. The minimum height of the base area is equivalent to the spray water height; this should be 300 mm. However, where the surface of the site is harder and level (slab covering), it should be raised to approximately 500 mm whilst if a 500 mm wide strip of gravel is laid (diameter 32 to 63 mm) at least 200 mm thick, and if the site slopes away from the base, it can be lowered to about 200 mm.

● The base area of the external wall must have a vertical seal on the surface or in the wall cross section and this seal must connect with the vertical seal of the underground external basement walls and with the horizontal wall seals in such a way that moisture bridges are avoided.

● Vertical seals in the base area can be made of the following:

– reinforced concrete at least 100 mm thick (see B 1.1.17);

– water-proof render (see B 1.1.7);

– or plastic films, which should be protected by a facing wall at least half a brick thick (for example of whole sand-lime bricks).

Sealing plasters, several layers of bituminous coatings or water-proof applied thin coatings, as well as brickwork made of clinker bricks and cement mortar (MG III) or small tiles applied with mortar should not be used as the sole form of seal.

● Materials not weather or moisture resistant (e.g. thermal insulation layers) should be applied to the inside of the vertical sealing layer. Only frost-resistant materials (for example clinker wall bricks) should be applied to the outside.

External basement wall
Structural wall and basement ceiling slab abutment

A large amount of the damage to external basement walls both in and above the ground was at least partially caused by the ground conditions of the site near the building. Base heights which were too low or back-filled soil which rose above the originally planned height of the site resulted in damp damage, even above existing horizontal cross section seals. Where surface water was drained into the site from a fairly large catchment area when the building was constructed on a slope or where adjacent courtyard surfaces incorrectly sloped towards the building, puddles formed in front of the wall and large quantities of water seeped into the excavated sites after back-filling, thus resulting in the accumulation of damp and corresponding damp damage to basement walls, which were generally not sealed against these loads.

Points for consideration

- The base area of the external wall is exposed to extreme water loads, the intensity of which is partly dependent on the shape of the adjacent site surface.

- The degree of spray water pressure and its level, and thus the position and level of the base area and the position of the upper horizontal wall seal, depend on the material and on the nature of the site surface (paved surface) (see B 2.2.4).

- A newly back-filled excavated site is subject to serious soil settling, particularly where it is back-filled with non-cohesive excavated soil or if it is not sufficiently compressed. This settling normally implies that the site surface will incline towards the wall thus promoting the formation of puddles. If the site is back-filled and compressed in layers with non-cohesive soil (soil replaced) the degree of settling is small. However, if the site is initially back-filled to above the required height by a suitable amount, this settling can be offset (see B 1.2.7).

- Large quantities of water traverse the site surface when the building is on a slope, particularly in heavy rain. Buildings cause this water to accumulate, thus resulting in the formation of puddles and, depending on the type of soil used for back-filling, in increased water in the excavated area. A similar effect is produced by large paved yard surfaces which increase the water pressure against the wall of the building either because of inadequate drainage or because they slope towards the building, and by the down pipe from roof surfaces, which are not connected to drainage pipes. This exposure to water running down a sloping surface can be prevented by means of open drainage channels or by drainage running parallel to the building, and, where the quantities of water involved are small, by sloping the surface of the site in the opposite direction.

Recommendations for the avoidance of defects

● The height of the final site surface along the external walls of the building must already be fixed at the design stage.

● The surface of the excavated site should where possible slope away from the external wall by about 5% once it has been back-filled and well compressed. If a fairly large amount of settling is to be expected, for example where wet, cohesive soil is back-filled, the site should be overfilled accordingly.

● Water being fed from the surrounding site to the external walls of the building should be avoided wherever possible, for example by trapping the water early in open drainage channels parallel to the building.

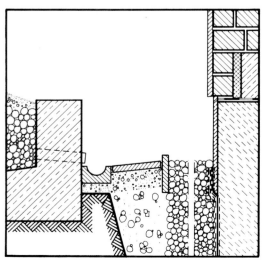

External basement wall
Structural wall and basement ceiling slab abutment

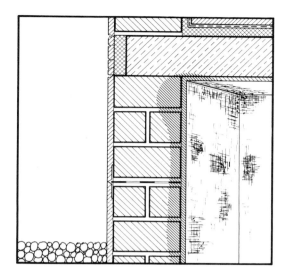

Damp damage caused by precipitated condensation appeared particularly on the inside surfaces of the basement external wall above ground level and in the corner where the wall meets with the reinforced concrete ceiling slab. The growth of black fungus, discolouration and peeling of paints and wallpapers as well as damage to plaster could be attributed to a lack of thermal insulation on the external edges of the reinforced concrete ceiling slabs or to a generally inadequate thermal insulation value of the external basement wall.

If thermal insulation on the external edges of the basement ceiling slab had been rendered over, wide cracks appeared in the joint between the thermal insulation material and the brickwork and there was also a network of fine cracks and the plaster had peeled away from the rotten thermal insulation boards, particularly where plaster-covered thermal insulation layers made of light wood fibre boards had been used in the base area. If the basement ceiling slab had been carried through to the outside and rendered over, or if different bricks had been laid in front of the external edges compared with the material used in the rest of the wall, cracks also appeared in the plaster at the joint with the wall surface.

Points for consideration

– Because of the high thermal conductivity of the reinforced concrete in the area where the basement ceiling slab abuts with the basement walls, basement ceilings represent cold bridges (depending on the materials used) with correspondingly low surface temperatures on the undersides. In addition, because of the reduced thermal transmission resistances caused by the low air movement in the corners, they also have the effect of reducing the temperature. In all examples, special thermal insulation measures are necessary.

– The thermal insulation value of conventional wall cross sections made of materials with a high thermal conductivity is at best approximately 0.65 m²K/W, in the case of bricks. Thus where the internal atmosphere of basement rooms contains a relatively high proportion of moisture (washing and drying rooms), and where the rooms are used for living purposes (playrooms, etc.) water may condense on the surface of the wall (see B 1.2.4). A further factor to be considered is the loss of heat through the external surfaces of the basement when heated. If basement rooms are not heated, a thermal insulation layer should be applied to the underside of the basement ceiling slab or the thermal insulation should be reinforced by a floating screed. The recommended minimum thermal transmission resistance amounts to 0.86 m²K/W (this is equivalent to a maximum thermal transmission coefficient of 0.93 W/m²K), but to obtain better conditions for habitable use of basements a maximum thermal transmission coefficient of less than 0.80 W/m²K (thermal transmission resistance of 1.04 m²K/W) is required.

– External rendering in the wall base area is mainly exposed to pressures from the substratum, particularly in the case of differences in movement at points where different materials or structural components meet. Render applied to thermal insulation materials is exposed to extreme temperature loads; expansion and shrinkage processes occur in vegetable thermal insulation materials as a result of varying moisture conditions, and these may place excessive pressures on the rendering. If the thermal insulation materials are constantly wet, they can be expected to rot and to damage the rendering.

External basement wall
Structural wall and ceiling slab abutment

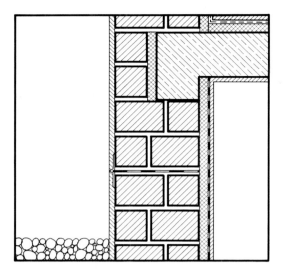

Recommendations for the avoidance of defects

● The thermal transmission resistance of the external basement walls must be suitable for the intended use of the basement rooms. In the case of heated rooms, part of which are in contact with the open air, the whole surface should have a thermal transmission resistance of at least 1.3 m²K/W ($k \leq$ 0.68 W/m²K) (see B 1.2.4).

● The external faces of the basement ceiling slab near the abutment with the external basement walls should be protected by thermal insulation layers with a thermal transmission resistance of at least 1.3 m²K/W ($k \leq$ 0.68 W/m²K).

● If it is not possible to apply external thermal insulation to the external face of the basement ceiling slab, 500 mm wide strips of thermal insulation material with a thermal transmission resistance of at least 0.65 m²K/W ($k \leq$ 1.24 W/m²K) should be applied to the underside of the ceiling slabs along the external walls.

● In the base area, the thermal insulation layers should be applied to the external basement walls or should be fitted in the wall cross section on the inside of the internal seal. The thermal insulation materials used should be moisture resistant and should have a high resistance to water vapour diffusion (e.g. polystyrene foam sheets). Where the thermal insulation material will not prevent the passage of water vapour (e.g. mineral fibre boards), a vapour barrier (e.g. laminated aluminium foil) should be applied to the inside.

Problem: Basement window and light well

Because of the growing trend not to use basements only as storage rooms, increased demands are being made in terms of damp protection and thermal insulation to the outer surfaces of the basement as well as in terms of the provision of better, natural lighting. If the site conditions do not allow the use made of the outside wall surfaces for direct glazing and the amount of space available does not permit sloping the site, enlarging the windows will also entail enlarging the light wells. These are becoming increasingly important and so, in contrast to current practice, more careful design and construction are necessary. This is particularly true in the case of light wells, which are exposed to an increased pressure from surface water and to accumulated water from the soil.

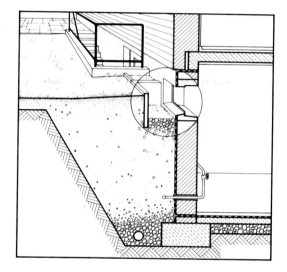

The damage observed in the area of the basement windows and basement light wells rarely affected the windows but mainly the light well construction and its connection with the external basement wall. Damp damage occurred to the inside of the basement particularly in the case of light wells mounted on stone corbels or consoles and whose walls were made of brickwork or prefabricated concrete sections. The causes of the damp penetration were mainly inadequate sealing measures to parts of the light well construction which had direct contact with the external brickwork of the basement and cracks where the light well walls connected with the basement brickwork.

The following pages describe defects in basement light wells and make recommendations for achieving defect-free light well construction.

External basement wall

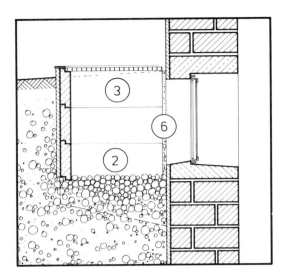

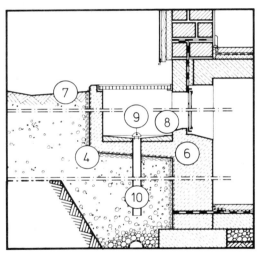

1 If basement light wells are to be permanently connected with the basement brickwork, they should be constructed with supporting side walls (B 2.3.2).

2 If the light well walls are to be built up or mounted on projecting floor slabs or cantilevers the amount of bending should be minimised when selecting the size of the supporting elements and to ensure that the supporting elements are firmly attached beneath the window soffit (see B 2.3.2).

3 If the supporting substratum is sufficiently compressed, light well walls whose foundations are separate from those of the building and which are set separately in front of the wall after the basement brickwork has been sealed are particularly recommended (see B 2.3.2).

4 Light wells permanently attached to the basement brickwork must be fully covered by the sealing measures applied to the external basement brickwork – particular attention should be paid to ensuring a careful seal at the joints with the basement brickwork, with the floor of the light well and, if applicable, with the supporting cantilevers (see B 2.3.3).

5 Where the soil conditions include water under pressure (groundwater, accumulated water) the basement windows and the light well should be built above the area of water under pressure. If this is not possible the light well should be continued as far as the basement base and should be included in the pressure-resistant seal of the basement floor and of the basement walls, or a drained ditch with separately sealed supporting walls should be constructed (see B 2.3.3).

6 If the light well walls can be expected to be exposed to more than soil moisture, efforts should be made to construct light wells that offer no risk of cracks appearing along the joint with the basement brickwork or that are not in contact with the basement brickwork (see B 2.3.3).

7 Site surfaces adjacent to the light well, especially where the building is on a slope, should be angled in such a way that surface water cannot enter the light well (see B 2.3.4).

8 The sill of the basement window should be a little above the bottom of the light well, which should slope away from the window (see B 2.3.4).

9 If exposed to large amounts of water, the bottom of the light well should be connected to the drainage system (gully) (see B 2.3.4).

10 In the case of drained excavated sites, the light well drainage can be connected directly to the adjoining land drainage by means of pipes (branch of the drains, stand pipe) or by means of a carefully constructed seepage layer with stable filtering properties (see B 2.3.4).

External basement wall
Basement window and light well

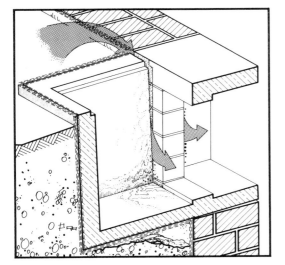

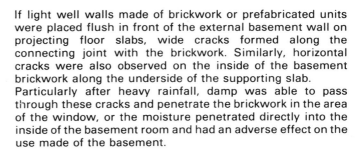

If light well walls made of brickwork or prefabricated units were placed flush in front of the external basement wall on projecting floor slabs, wide cracks formed along the connecting joint with the brickwork. Similarly, horizontal cracks were also observed on the inside of the basement brickwork along the underside of the supporting slab.

Particularly after heavy rainfall, damp was able to pass through these cracks and penetrate the brickwork in the area of the window, or the moisture penetrated directly into the inside of the basement room and had an adverse effect on the use made of the basement.

Points for consideration

– Projecting light well floor slabs and cantilevers cannot be supported at any required height by light well walls or precast units without sustaining damage – even though the converse often appears in product information sheets from manufacturers. Even with shallow light wells, the supporting structures can be damaged so severely that wide cracks appear in the joint between the light well and the external basement wall. The back-filled soil cannot be expected to have the effect of carrying away this load since it is as a rule not possible to compress this soil sufficiently below the light wells.

– Because of the unfavourable position of the attachment of the supporting elements directly beneath or beside the basement window opening, slight twisting of the abutment is possible, and this will cause cracks in the area of the abutment.

These problems do not occur in light wells whose side walls tie into the brickwork for their full height.

– As a rule, it is not so much the cracks themselves as the damp penetration caused by the cracks that results in damage. This damage in the form of cracks should therefore be prevented, especially as the light well is heavily exposed to surface and seepage water.

– If the light well has been constructed at a later date, independently of the external basement walls and if it is possible to apply a permanent seal to the external basement wall without including the light well surfaces, the connecting joint with the external brickwork of the basement does not provide a sealing function at this point and can then result in severe damp problems.

Recommendations for the avoidance of defects

● If basement light wells are to be permanently connected with the basement brickwork, they should be constructed with supporting side walls.

● If the light well walls are to be built up or mounted on projecting floor slabs or cantilevers, the amount of bending should be minimised when selecting the size of the supporting elements and to ensure that the supporting elements are firmly attached beneath the window soffit.

● If the supporting substratum is sufficiently compressed, light well walls, whose foundations are separate from those of the building and which are set separately in front of the wall after the basement brickwork has been sealed are particularly recommended.

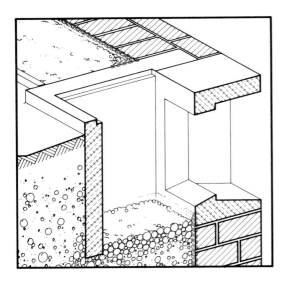

External basement wall
Basement window and light well

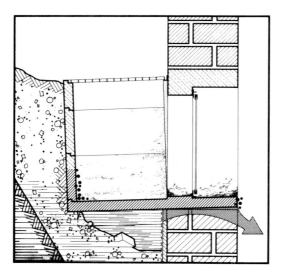

Damp damage to the inside of the basement brickwork around the basement window was observed frequently if light wells made of brickwork or prefabricated concrete sections were connected directly to the brickwork whilst the light well surfaces were not sealed or were inadequately sealed. Inadequate sealing was particularly common under light well floors and on the cantilevers supporting light well constructions. If the joints of light wells' supporting slabs or the connecting joints between the light well side walls and the basement brickwork were inadequately sealed or if they were not water-proof despite attempts at sealing them with water-proof rendering, compounds or bituminous coatings which had pulled away, the water seeped through the joints into the light well, from where it also penetrated the inside of the adjacent basement brickwork.

Damp damage occurred in particular where the light well was exposed to large quantities of water in the form of accumulated water in a poorly drained excavated site on a slope or one which had cohesive soil.

Points for consideration

- Light wells connected with the basement brickwork became damp and the damp penetrated the adjacent basement brickwork if the surfaces of the light well which was connected with the ground (such as the basement brickwork itself) were not adequately sealed.

- There are many opportunities for making mistakes when applying a complete seal to the basement light well because of the many angled surfaces. It is very expensive and difficult to produce a pressure resistant seal for a light well floor because of the pressure and compression needed. In the case of heavier water pressures (accumulated water, groundwater) therefore, the light well represents a particular defect in the sealing of the external basement wall.

- If the light well walls are connected flush to the external basement brickwork, it is easy for a crack to appear along the connecting joint. This crack will be produced by the bending of the supporting elements of the light well and by the settling movements of the building. Using acceptable constructional methods, it is impossible to bridge the crack formed in this way with sealing layers.

Recommendations for the avoidance of defects

● Light shafts permanently attached to the basement brickwork must be fully covered by the sealing measures applied to the external basement brickwork. Particular attention should be paid to ensuring a careful seal at the joints with the basement brickwork, with the floor of the light well and, if applicable, with the supporting cantilevers.

● If the light shaft walls can be expected to be exposed to more than soil moisture, efforts should be made to construct light shafts that offer no risk of cracks appearing along the joint with the basement brickwork or that are not in direct contact with the basement brickwork (see B 2.3.2).

● Where the building is exposed to water under pressure (groundwater, accumulated water), the basement windows and the light well should be built above the area of water under pressure. If this is not possible the light well should be continued as far as the basement base and should be included in the pressure-resistant seal of the basement floor and of the basement walls, or a drained ditch with separately sealed supporting walls should be constructed.

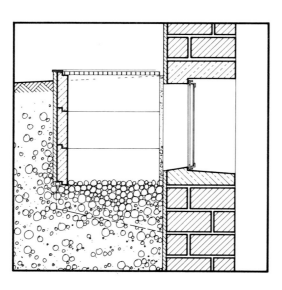

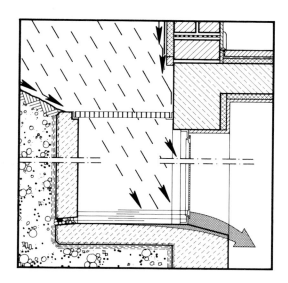

If the light well had not been provided with drainage facilities, or if drainage openings were made too small and became blocked, water collected in the light well. If the site sloped down towards the light well, surface water entered the room, particularly if the basement window parapet was not built higher than the floor of the light well.

Points for consideration

– Considerable quantities of water were able to enter the light well, especially on sides of the building exposed to heavy rain and when the site sloped down towards the building.

– A sufficiently high threshold between the sill of the basement window and the floor of the light well provides initial protection against the entry of water.

– A hole in the base slab alone does not guarantee reliable drainage since, depending on the type of material used to back-fill the excavated site, the capacity of the soil to absorb water can be very low. Because of this, therefore, uncontrollable seepage of the water collecting in the light well involves many risks (particularly where large quantities of water are present – large light wells, sides of the building exposed to heavy rain, building constructed on a slope), and changing this during the course of any subsequent improvements will involve high costs.

Recommendations for the avoidance of defects

● Site surfaces adjacent to the light well, especially where the building is on a slope, should be angled in such a way that surface water cannot enter the light well.

● The sill of the basement window should be at least 100 mm above the bottom of the light well, which should slope away from the window to the outlet.

● In the case of large light wells, and heavy exposure to water, the light well floor should be connected to the drainage system (via a gulley).

● In the case of drained excavated sites, the light well drainage can be connected to the adjoining land drainage by means of pipes (branch of the drains, stand pipe) or by means of a carefully constructed seepage layer with stable filtering properties.

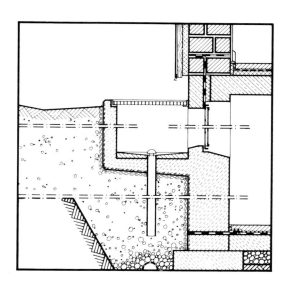

Problem: External basement stairs

External basement stairs are normally independent of the basement structure, so the damage caused by surface cracking and water is of minor importance. However, in the case of basement stairs, easily avoided failures should not be tolerated. Moreover research into structural failure has shown that damage and inadequacies connected with the basement stairs can frequently cause further problems in adjacent basement rooms and this damage may be considerably more serious than the initial failure. This is therefore a further reason why greater attention should be paid to the designing of the basement stairs.

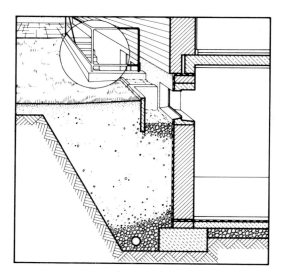

The damage which occurred mainly consisted of damp penetration – however cracking was also observed, mainly at points where stairs parallel to the external basement wall joined the structural wall of the house. The design and construction difficulties encountered in the case of external basement stairs are typical problems associated with joining and sealing external structural components connected to the building as well as typical problems of stability and of the protection of supporting walls against damp. To this extent, therefore, the defects and the recommendations for avoiding them which are outlined on the following pages will not only be of interest in the design and construction of trouble-free external basement stairs.

External basement wall

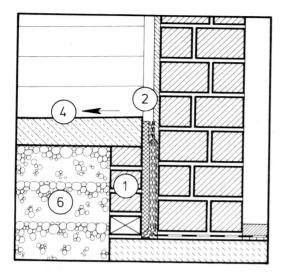

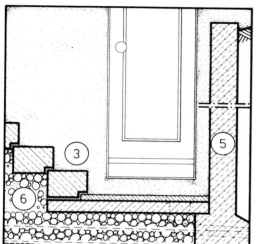

1 The external basement stairs should have a separate foundation which is independent from that of the external basement wall and should be separated from the building by a joint, particularly where surface seals are applied to the outside of the basement walls or if the building can be expected to settle. If the basement and the stairs are made of water-proof concrete then a permanent connection between the stairs and the wall is possible if the stairs are built in the form of a reinforced cantilever or if they are cast, together with the building, on a rigid slab (see B 2.4.2).

2 In examples of basement walls with surface seals applied to the outside, the basement stairs should where possible not be built until after the sealing work has been completed. The stairs should be placed on a foundation separate from the building and should be separated from the wall by a joint at least 20 mm wide (see B 2.4.3).

3 Surface materials which are suitable for ground level areas should be applied to the wall surface at the top of the stair and must be carried as a strip at least 300 mm wide (see B 2.4.3).

4 The surface of the stairs should slope slightly away from the basement wall (approximately 1%) (see B 2.4.3).

5 Wherever possible, basement stairs and retaining walls which are built in the ground should be made from water-proof concrete. In addition, the size of these components must be able to resist the ground pressure to be expected (see B 2.4.4).

6 Particular attention should be paid to careful back-filling of the excavated site and hollow spaces in the area of the basement stairs with non-cohesive filling material (e.g. gravel). In the case of cohesive soils, designed drainage measures should, if necessary, include the area around the basement stairs especially if no action has been taken to provide protection against water under pressure (see B 2.4.4).

External basement wall
External basement stairs

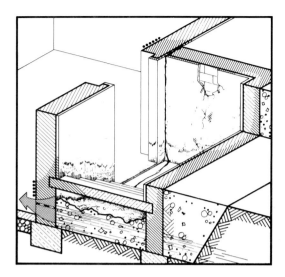

If one wall of the stair well was built into the external basement wall and the other wall was supported by a retaining wall built separately on a strip foundation, cracking damage was observed at points where the stairs and the surrounding retaining walls connected with the basement wall brickwork.

Water was able to enter these cracks and penetrate the inside of the building through the basement wall brickwork. This restricted the use of the rooms, particularly where the water pressure was heavy, if the building was on a slope, and this restriction was generally more disruptive than the initial cracking damage.

Points for consideration

- If the external basement stairs are either built on foundations which are completely separate from the building or if they are supported partly by the basement wall and partly by a supporting wall with separate foundations, then differences in settlement are to be expected, since the building foundations settle more than the stairs because of the heavier weight on the soil. If under such conditions the stairs are connected rigidly to the basement wall cracking will occur as a result of the traction and shearing stresses which take place because of settlement differences. This is particularly true in the case of brick walls and non-reinforced concrete walls along the area where the stairs abut.

- Uncontrolled cracking can be prevented if the stairs are built on a separate foundation from the external basement wall and if they are separated from the building by a joint.

- The cracking can also be prevented if the basement stairs and the basement wall are supported on a common rigid slab and if they are firmly connected to the building in the form of a continuous reinforced concrete basement construction (e.g. in the form of a reinforced cantilever construction).

- In the case of basements with external surface seals, all stairs which are connected with the building must be included in the sealing measures. Since as an external structural component all parts of the basement stairs are exposed to damp from rain and from soil moisture, all surfaces of the angled stair must be sealed and this can rarely be perfect if surface sealing measures (e.g. bituminous coatings, sealing compounds, waterproof rendering, building papers) are used. The method whereby the stairs are rigidly connected to the building thus offers the best solution in the case of water-proof concrete basement constructions.

- The risk of damage caused by uncontrolled cracking can possibly be accepted in the most economical type of design in which a staircase is supported on one side in the basement wall without a joint, provided the soil, foundation and load conditions imply that only slight settlement is to be expected, the water pressure is slight (soil moisture), and provided the basement is protected by internal sealing. In such examples any slight, negligible cracking that might occur would not result in any damp penetration.

Recommendations for the avoidance of defects

● The external basement stairs should have a separate foundation independent from that of the external basement wall and should be separated from the building by a joint, particularly where surface seals are applied to the outside of the basement walls or if the building can be expected to settle.

● If the basement and the stairs are made of water-proof concrete a permanent connection between the stairs and the wall is possible if the stairs are built in the form of a reinforced cantilever or if they are cast, together with the building, on a rigid slab.

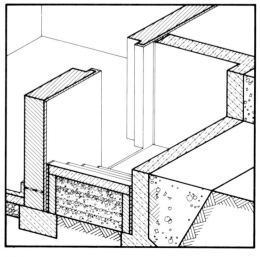

External basement wall
External basement stairs

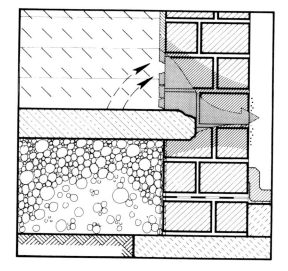

In the case of basement walls with surface seals – usually in the form of bituminous coatings – damp damage appeared in the area of the basement stairs particularly if the stairs were supported as precast units in the basement wall. Where this type of construction was used, cracks were noticed in the joint. The bituminous coating which was applied after the stairs had been built was incomplete beneath the stairs and external rendering that had been continued to the surface of the stairs had peeled away as a result of frost.

Damp penetration was particularly heavy if the surface of the stairs were angled towards the basement wall.

Points for consideration

- If in the case of basement walls with external surface seals (bituminous coatings, waterproof renders, sealing compounds, building papers or felt) the stairs and the supporting walls tie in with the basement walls, the joint at the point where they are tied in is broken. If the structural components which are tied in are themselves not water-proof moisture can penetrate the basement wall under capillary action. In addition, because of major settlement of the building, a crack frequently appears at the abutment and this results in heavy damp penetration, especially when the stairs had angled towards the building as a result of settlement.

- If the wall area below the stairs was not sealed until after the stairs had been built, overall application of the sealing materials is often no longer possible in these inaccessible places.

- If the vertical wall seal was interrupted, cracks in the abutment and the difficulties encountered in providing external sealing measures can be prevented if the basement stairs are not constructed until the wall seal has been completed and if they are independent of the basement wall, e.g. if they are supported on a sub-wall or are separated from the wall surface by a joint.

- In the case of basements made of water-proof concrete and internal sealing measures, the way in which the basement stairs are connected to the building has no effect on the water-proofness of the external basement wall.

- The area of wall above the surface of the stairs is exposed to spray water and in some instances to standing water. In this area, therefore, materials suitable for ground level areas should be used in order to prevent damp penetration of the wall and sometimes frost damage (see B 2.1).

Recommendations for the avoidance of defects

● In examples of basement walls with surface seals applied to the outside, the basement stairs should where possible not be built until after the sealing work has been completed. The stairs should be placed on a foundation separate from the building and should be separated from the wall by a joint at least 20 mm wide.

● Surface materials suitable for ground level areas should be applied to the wall surface at the top of the stair and must be carried as a strip at least 300 mm wide.

● The surface of the stairs should slope slightly away from the basement wall (approximately 1%).

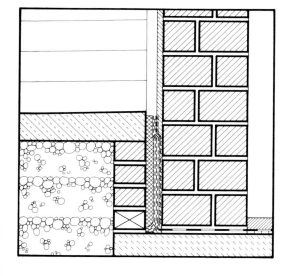

External basement wall
External basement stairs

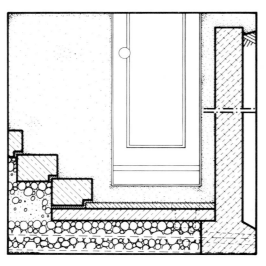

Harmful damp penetration in the area of the stairs was noticed when precast concrete steps were not closely joined to one another so that water could seep out from between them or if the stair supporting wall was not adequately sealed against exposure to water from the soil so that a rendered coating applied to the inside became damp, blistered and peeled off. If a horizontal damp-proof course made of bituminous coated felt was inserted into the basement supporting walls above the floor slab in order to protect the walls against rising damp, the wall was displaced laterally on a level with the horizontal damp-proof course when the site was back-filled.

Much of the damp damage in the area of the stairs was aggravated by faults that occurred during back-filling. This was particularly true in the area beneath the stairs, where there were frequently hollow cavities which filled with seepage water, especially in examples of stairs built on the side of a slope, so that the basement wall and the stairs were exposed to water.

Points for consideration

– Although the demands made on the external basement stairs, which is a structural component of minor importance, are not particularly heavy, the damp penetration they cause, which will lead to restrictions in the use of the basement rooms, should be avoided.

– Supporting walls made of brick, which are in the ground, suffered various problems. If damp penetration, blistering and peeling as a result of frost of the surface layers on the inside of the wall are to be avoided, those parts of the brick walls in the ground must be sealed against the type of water pressure to which they are exposed. In order to prevent rising damp (from the foundations and from water standing on the base of the stairs), a horizontal damp-proof course must be inserted in the cross section about 100–150 mm above the floor slab. Since the load from the section of supporting wall above the horizontal seal is relatively small, low friction resistances can be expected on a level with the horizontal damp-proof course. The lateral pressure of the back filling material may thus cause the wall to be displaced in this area. In the case of supporting walls made of water-proof concrete it is considerably easier to achieve damp protection and to absorb the lateral pressure from the ground.

– The area beneath the basement stairs is inaccessible. In many cases, therefore, it remains partially unfilled, particularly when the excavated site is back-filled by machine. Similarly, the back-filling material cannot be compressed. The seepage water that collects in the hollow cavities can place a heavier pressure on the basement wall and stairs in the form of accumulated water, especially in cohesive soils and where the back-filling material contains building debris. If the space is carefully back-filled by hand with non-cohesive material (e.g. gravel), or if this space is drained by the drainage system in cohesive soils, then this exposure to accumulated water can be avoided.

Recommendations for the avoidance of defects

● Wherever possible, basement stairs and supporting walls which are built in the ground should be made from water-proof concrete. In addition, the size of these structural components must be able to resist the ground pressure to be expected.

● Particular attention should be paid to careful back-filling of the excavated site and hollow spaces in the area of the basement stairs with non-cohesive filling material (e.g. gravel). In the case of cohesive soils, the designed drainage measures should, if necessary, include the area around the basement stairs especially if no action has been taken to provide protection against water under pressure.

Problem: Joints and openings in the basement wall

Two further areas of the external basement wall seals appeared to cause problems, although less commonly: these were firstly joints, particularly joints between house party walls, for example in the construction of terraced houses or when building on to existing buildings and secondly openings, such as those carrying pipes for services. Contrary to current popular opinion, the empty space left between cavity walls in order to avoid sound transmission or settlement damage can be expected to suffer exposure to accumulated water from the sides or from below, depending on the prevailing water conditions in the surrounding soil, unless a comprehensive seal or effective drainage is provided. In contrast, areas where openings occur are generally in a clear section of wall and are thus directly exposed to seepage or accumulated water from the excavated site. They must therefore be suitably linked to the vertical wall seal. Similarly, possible movement and vibration (e.g. during back-filling) of pipe runs which pass through the wall should also be taken into account in design and execution.

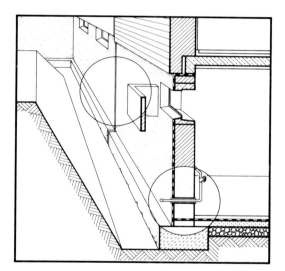

At openings and joints between house party walls only damp penetration damage was observed and this resulted in more or less extensive secondary damage. The reason for the damage was the failure to observe the above requirements.

The fact that in these linked areas no damage occurred when tanking constructions were undertaken leads one to assume that these points of detail were designed and constructed more carefully in full knowledge of the possible follow-up damage.

The typical, recurring damage, the processes that cause this damage and recommendations for avoiding it are discussed below.

External basement wall

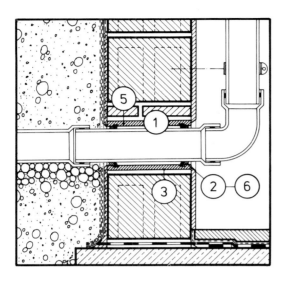

1 The position and number of pipe ducts and the openings in external walls needed to carry them should be determined as early as possible at the design stage. In particular, pipe ducts should wherever possible be avoided with seals using building papers, felts or tanking constructions, or at least, several individual pipes should be combined to form a single duct (see B 2.5.2).

2 The openings must be constructed in such a way that it is possible to obtain a tight connection with the sealing layer and to absorb tolerances and movement (see B 2.5.2).

3 Pipes should be covered with protective sleeves where they pass through the external wall and a designed connection must be made between these pipes and the vertical wall seal in terms of the prevailing external pressures and the sealing method selected (see B 2.5.2).

4 In the case of exposure to accumulated water and where the sealing method uses building papers, a steel protective pipe, for example, should be selected. As far as the construction is concerned, detailed specifications (connection to the building paper, distance between the flange and pipe sockets, etc.) should be noted (see B 2.5.2).

5 When sealing the wall with bituminous coatings, waterproof rendering or sealing compounds, the outer sleeve should be made of a material which is the same as or similar to the wall material (e.g. concrete, earthenware, asbestos cement) (see B 2.5.2).

6 The size of the internal diameter of the outer sleeve must be such that the pipes have sufficient play for the movements which are to be expected and it must be such that a permanently tight connection of the pipes is possible (e.g. with elastic sealing rings or mastic joint sealing compounds) and if necessary so that the pipe can be surrounded with thermal insulation material (see B 2.5.2).

7 If the accumulation of water between house party walls cannot be reliably avoided by means of effective drainage and sealing measures, and if only one of the walls can be sealed from the outside without damage, the wall to be built afterwards should either be made from water-proof concrete or an internal seal should be used (see B 2.5.3).

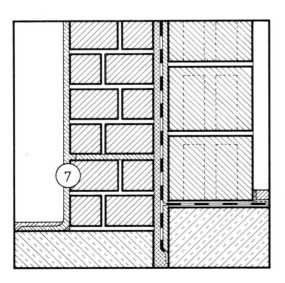

External basement wall
Joints and openings in basement walls

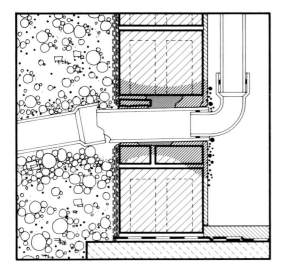

In the area of openings in external basement walls to carry pipes providing the necessary services for the building (electricity, water, gas, telephone) damp penetration was particularly common if, because of the site situation (slope) or because of the soil and groundwater conditions, the openings were exposed to a water pressure other than that from soil moisture. The damage was caused because there was no seal around the pipe openings or because the sealing method selected was unsuitable and did not allow for movement in the pipes. If the seals were made of rigid material (e.g. mortar), cracks appeared directly in the contact surfaces of the pipes that passed through the wall without any protective covering or outer sleeve. In some cases, soft seals (mastic compounds) peeled away from their edges and were unable to withstand constant exposure to moisture. Besides small damp patches in the immediate vicinity of the openings there were also – depending on the size of the unsealed opening and on the external water pressure – cases of more extensive damp penetration, some of which were penetrated by bituminous compounds from the external seal and this resulted in damage to the internal plaster, paintwork and wallpaper, and partially restricted the usefulness of the rooms.

Points for consideration

- Any opening in the basement wall that is below ground level is an interruption in the sealing layer. It thus also represents an area of additional risk of damp penetration and this risk increases as the number of openings multiplies, and the way in which they are to be sealed must be considered with care both in their design and execution, the heavier the water pressure from the soil.

- Metal pipes in particular are prone to considerable movement, especially lengthwise, because of their high thermal expansion in temperature fluctuations. Moreover, sections of pipe outside the building that are in the ground (including cable conduits) can settle considerably (especially where there is a soft, unstable base) when they are under pressure that excessive loads can be placed on the rigid pipe passing through the wall as a result of displacement, tilting or twisting of the pipe. This will result in the peeling of the render and sealing layers or the formation of cracks that will allow water to penetrate. Therefore, in all areas where pipes pass through external basement walls, particular care should be taken to ensure that all the vertical wall seals are tightly connected to the area of the opening and water bridges are prevented and that the pipes can move without causing damage.

- Where sealing layers are connected directly with pipes, possible movements are transmitted in full to the seal and can thus cause cracks and failures in the sealing layer if the pressure is excessive. Protective packing or outer sleeves surrounding the pipe, which are fitted into the wall, reduce the danger of this. Provided that there is sufficient clearance, they allow a certain amount of room for movement and in some cases make it possible to cancel out tolerances, to lay pipes at a slight angle or to replace pipes that may have become damaged, without damaging the wall seal.

- Brittle materials (e.g. water-proof mortars) do not provide for movements without damage. They are therefore not suitable for use as sealing materials around pipes. When using mastic jointing compounds it is important to remember that the sealing effect depends on their permanent elasticity, and on their resistance to water, chemicals and aging and particularly on the adhesion of the sealing compound to the edges with which it is in contact.

119

External basement wall
Joints and openings in basement walls

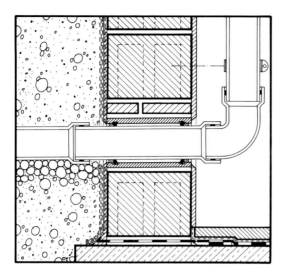

Oily preliminary adhesion coatings or those which are soiled with dust or bitumen or which are water-absorbent do not provide a suitable substrate for adhesion.

Recommendations for the avoidance of defects

● The position and number of pipe ducts and the openings in external walls needed to carry them should be determined as early as possible at the design stage. In particular, pipe ducts should wherever possible be avoided when seals using building papers, felts or tanking constructions are used, or at least, several individual pipes should be combined to form a single duct.

● The openings must be constructed in such a way that it is possible to obtain a tight connection with the sealing layer and to absorb tolerances and movement.

● Pipes should have protective sleeves where they pass through the external wall and a designed connection must be made between these pipes and the vertical wall seal in terms of the prevailing external pressures and the sealing method selected.

● In the case of exposure to accumulated water and where the sealing method uses building paper a steel protective covering, for example, should be selected. As far as the construction is concerned, detailed specification (connection to the seal, distance between the flange and pipe sockets, etc.) should be noted.

● When sealing the wall with bituminous coatings, water-proof rendering or sealing compounds, the outer sleeve should be made of a material that is the same as or similar to the wall material (e.g. concrete, earthenware, asbestos cement).

● The size of the internal diameter of the outer sleeve must be such that the pipes have sufficient play for the movements which are to be expected and must be such that a permanently tight connection of the pipes is possible (e.g. with elastic sealing rings or mastic joint sealing compounds).

External basement wall
Joints and openings in basement walls

In the cavity dividing walls of terraced houses and in houses built on later, damp damage was visible on the inside and generally spread across the whole length of the wall above floor level. In all cases, the building, soil or water conditions were such that water was able to rise to various heights in the cavity between the wall leaves, some of which were not completely sealed; the height of the area affected by damp depended on the height of the accumulated water and on the position of the horizontal damp-proof course.

In some cases, particularly where additional structures had been built on at a different height, the adjoining land drain pipes were broken at this point, thus resulting in the increased accumulation of water which could then seep into the unsealed cavity.

Points for consideration

– In order to take into account varying degrees of settlement in the case of long buildings, as well as varying building loads or variations in the ground conditions, and in order to achieve increased sound insulation, as well as for other constructional reasons, double leaf cavity walls which extend as far as the foundations are used, particularly in terraced houses. The narrow cavity between the walls means that an external surface seal cannot be applied to at least one of the two walls and because of this, water that enters the cavity from the side or from below can penetrate the wall which has not been sealed. The area prone to damp is particularly extensive if the water is able to accumulate above the lower horizontal damp-proof course of the wall.

– Comprehensive joint sealing strips made of metal or plastic which are inserted into the wall cross section or which are linked with the horizontal damp-proof course in the lower area and which are joined with the vertical wall seal on the sides are suitable for sealing the edges of the wall cavity but they must be designed and executed with the utmost care. Leaks can easily occur as a result of failures, particularly in areas where vertical and horizontal seals join and if there are projections in the wall. Moreover it is not always possible to prevent water from circulating and percolating behind the edges of the sealing strips.

– Similarly, the use of soft joint sealing compounds to seal horizontal sealing layers is not absolutely reliable. Mastic sealing compounds can be used in the vertical joint area, but this is only functional if no large amount of settlement is to be expected and if the vertical wall seal consists of water-proof rendering or sealing compounds which ensure a perfect joint edge and which make suitable preparation with water-proof adhesive priming coats possible.

– Any interruption in the adjoining land drainage running around the building will allow additional water to penetrate the wall cavity. The drainage should therefore take the form of a closed circuit – even in the case of buildings at different heights (if necessary by installing shafts or reinforced seepage packings at the points where the buildings are of different heights (see A 1.1.2).

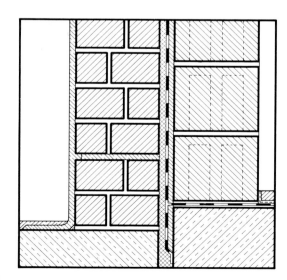

Recommendations for the avoidance of defects

● If the accumulation of water between house party walls cannot be reliably avoided by means of effective drainage and by seals in the vertical joint area, and if it is possible to seal only one of the walls from the outside, then the wall to be built afterwards should either be made from water-proof concrete (see B 1.1.15–B 1.1.17) with a suitable connection to the basement floor slab (see C 2.1.2), or an internal seal should be used (see B 1.1.11 and B 1.1.12).

121

External basement wall
Points of detail

Base and basement ceiling abutment

Lufsky, Karl: Bauwerksabdichtung, Bitumen und Kunststoffe in der Abdichtungstechnik. 3. Auflage, B. G. Teubner, Stuttgart 1975.

Reichert, Hubert: Paradontose an Häusern – vermeidbar? In: Deutsche Bauzeitschrift (DBZ), Heft 12/1975, Seite 1457–1458.

Reichert, Hubert: Sperrschicht und Dichtschicht im Hochbau. Verlagsgesellschaft Rudolf Müller, Köln 1974.

DIN 1053 – Blatt 1: Mauerwerk, Berechnung und Ausführung. November 1974.

DIN 4117: Abdichtung von Bauwerken gegen Bodenfeuchtigkeit, Richtlinien für die Ausführung. November 1960.

DIN 18195 – Teil 4 (Entwurf): Bauwerksabdichtungen, Abdichtungen gegen Bodenfeuchtigkeit, Ausführung und Bemessung. November 1977.

Basement window and light well, external basement stairs

Behringer, Anton C.; Rek, Franz; Haeberle, Kurt: Das neue Maurerbuch. 10. Auflage, Otto Maier, Ravensburg 1966.

Frick; Knöll; Neumann: Baukonstruktionslehre. Teil 1, 25. Auflage, B. G. Teubner, Stuttgart 1975.

Schild, Erich: Untersuchung der Bauschäden an Kellern, Dränagen und Gründungen. Forum Fortbildung Bau, Heft 8, Forum-Verlag, Stuttgart 1977, Seite 49–67.

Schmitt, Heinrich: Hochbaukonstruktion. 6. Auflage, Vieweg, Braunschweig und Wiesbaden 1977.

Joints and openings in basement walls

Blöchl, Otto: Fehlerquellen bei der Hausentwässerung. In: deutsche bauzeitung (db), Heft 6/1976, Seite 48–50.

Grunau, Edvard B.: Fugen im Hochbau. Verlagsgesellschaft Rudolf Müller, Köln 1973.

Volger, Karl: Haustechnik. 4. Auflage, B. G. Teubner, Stuttgart 1971.

DIN 1986 – Blatt 1: Entwässerungsanlagen für Gebäude- und Grundstücke, Technische Bestimmungen für den Bau. August 1974.

DIN 1988: Trinkwasser-Leitungsanlagen in Grundstücken, Technische Bestimmungen für Bau und Betrieb. Januar 1962.

DIN 4031: Wasserdruckhaltende bituminöse Abdichtungen für Bauwerke, Richtlinien für Bemessung und Ausführung. November 1959.

DIN 18195 – Teil 6 (Entwurf): Bauwerksabdichtungen, Abdichtungen gegen von außen drückendes Wasser, Ausführung und Bemessung. November 1977.

DIN 4122: Abdichtung von Bauwerken gegen nichtdrückendes Oberflächenwasser und Sickerwasser mit bituminösen Stoffen, Metallbändern und Kunststoff-Folien, Richtlinien. Juli 1968.

DIN 18012 – VOB, Teil C: Hausanschlußraum, Bautechnische Richtlinien. Juni 1964.

DIN 18307 – VOB, Teil C: Gas- und Wasserleitungsarbeiten im Erdreich. Dezember 1973.

Problem: Sequence of layers and individual layers

The traditional basement floor constructions which are still mainly used today often no longer meet the requirements placed on them by the uses made of the basement rooms. Rooms whose floors have either no sealing measures whatsoever, or which have only coarse filling beneath the floor slab and cementitious screeds in order to interrupt capillary action are frequently used not only for storing goods that are not sensitive to moisture, but are fitted out as guest rooms, playrooms, hobby rooms, etc. and are provided with suitable floor coverings (carpets, parquet flooring, floating screeds). This results in damp damage to the covering layers and damp, musty conditions inside the rooms which, unless it is decided to alter the use made of the rooms, necessitate expensive additional work.

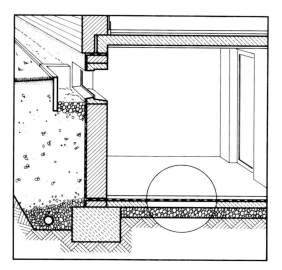

Formation of puddles and flooding represent a second group of damage which year after year make it impossible to use basements. This is the result of inadequate sealing of the basement floor because exposure to water under pressure was not adequately determined and mainly because as the deepest part of the enclosure of the building, the basement floor is most heavily exposed to groundwater.

In contrast, damage to basement floors, which were built as part of tanking construction resistant to water under pressure, was very rare, since the available provisions and design information are aimed specifically at a high degree of safety and since the designing and construction are carried out in full awareness of the consequences of inadequacies.

The recommendations made on the following pages therefore concentrate particularly on floor constructions beneath rooms that are expected to remain dry and on methods of construction that can withstand temporary exposure to water under pressure. Similarly, many aspects of the recommendations made here for floor constructions can also be applied to buildings that do not have basements.

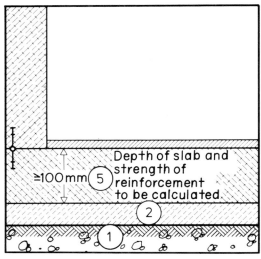

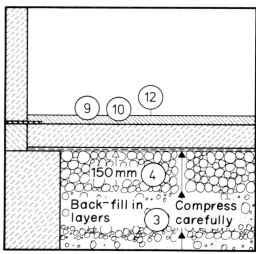

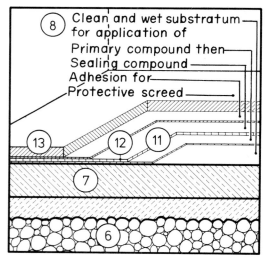

1 Attention should be paid to ensuring that the substratum of the basement floor slab is level and well compacted (see C 1.1.2 and C 1.1.3).

2 The most reliable way of ensuring that the substratum is level is to use a smoothing layer of lean concrete about 80 mm thick (see C 1.1.2).

3 Where the site to be filled is deep, non-cohesive back-filling material should be introduced in layers about 300–400 mm thick and should be compacted (see C 1.1.2).

4 If no sealing layer is going to be used in the basement floor where it is exposed to soil moisture because the basement is not to be used for any major purpose, a coarse filling at least 150 mm deep should be laid beneath the floor slab (see C 1.1.2 and C 1.1.3).

5 In the case of exposure to water under pressure, the floor slab should take the form of a rigid reinforced concrete slab, the size of which should be determined by calculation. Floor slabs not exposed to water under pressure should be at least 100 mm thick (see C 1.1.3).

6 Water-proof screeds and compounds should only be used if the prolonged accumulation of seepage water is impossible, for example because of the provision of surface drainage (see C 1.1.4–C 1.1.6).

7 Water-proof screeds and sealing compounds should only be applied to level, non-reinforced, jointless floor slabs (see C 1.1.5 and C 1.1.6).

8 Before water-proof screeds or sealing compounds are applied, the substratum should be free of loose particles and it should be wetted and covered with a cement-sand slurry or with a priming slurry (see C 1.1.5 and C 1.1.6).

9 Particular attention should be paid to ensuring that the composition of the cement screed is correct in terms of grain composition, cement content and water-cement value. Additives used according to the manufacturer's instructions may improve water-proofness (see C 1.1.5).

10 The screed should be applied in one operation to a thickness of about 30 mm, should be compacted and should be smoothed without excessive rubbing and without the use of gauges (C 1.1.5).

11 Sealing compounds should be applied in several continuous layers. The minimum coverage amounts specified by the manufacturer should be observed or, if necessary, exceeded. The compound should be distributed and compacted with a float (see C 1.1.6).

12 Until they have fully hardened, waterproof screeds and sealing compounds should be protected against uneven or excessively fast drying and if necessary should be wetted (see C 1.1.5 and C 1.1.6).

13 Immediately after they have fully hardened, sealing compounds should be protected against damage by being covered with a protective coating (see C 1.1.6).

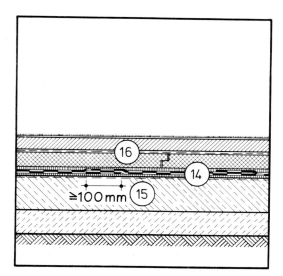

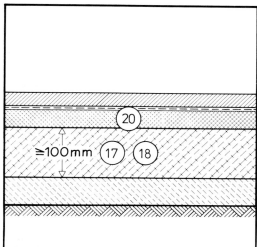

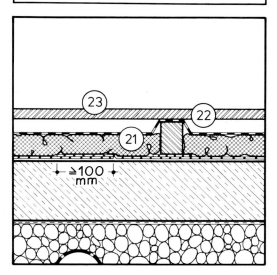

14 If the basement floor has to be very dry, sealing coats should generally be applied to the floor slab, provided possible exposure to water under pressure does not necessitate the use of tanking construction with compacted sealing courses or one which is made from water-proof concrete (the tanking construction to be produced by specialist firms) (see C 1.1.4, C 1.1.7 and C 1.1.8).

15 Double skin seals with offset seams glued to the whole surface and covered with a protective coating are preferable to single layer seals. Particular attention should be maintained in obtaining a sufficiently wide overlap ($\geq$ 100 mm) and to ensuring that adhesion is correct (see C 1.1.7).

16 Immediately after laying, building paper or felt seals should be provided with a protective covering, which should be separated from the seal by a sheet of foil (see C 1.1.7).

17 Water-proof concrete floor slabs used as part of water-proof concrete tanking which are resistant to water under pressure should only be manufactured by specialist firms. Their size should be such that cracking is limited to a width of less than 0.1 mm. In other cases, water-proof concrete floor slabs should only be used to provide protection against soil moisture and non-accumulating seepage water (see C 1.1.8).

18 Particular attention should be paid to the composition of the water-proof concrete in terms of material and grain composition of the additive, the proportion of cement and to the water cement value. Sealing agents can be added, but their suitability must be proven (see C 1.1.8).

19 Joints in the work should be avoided wherever possible. If, however, they are unavoidable, they should be carefully designed. The water-proof concrete should be introduced in such a way that separation is prevented and a good seal is produced (see C 1.1.8).

20 The floors of living rooms which are below ground level must have a thermal transmission resistance of at least 0.86 m^2K/W by the use of water-proof thermal insulation layers (see C 1.1.9 and C 1.1.10).

21 Thermal insulation layers and moisture-sensitive floor coverings should only be fitted on top of high grade seals (see C 1.1.10 and C 1.1.11).

22 The upper surface of insulation layers which have a low resistance to water vapour should be protected by means of a vapour barrier (see C 1.1.10).

23 Non-ventilated floor coverings should not be applied to moisture-sensitive underlays, or to underlays liable to rot (C 1.1.11).

Basement floor
Sequence of layers and individual layers

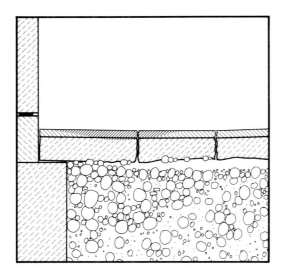

A number of examples of cracking damage in the floor slab and most of the damp damage which this produces can be attributed to faults in the substratum of the basement floor. In most cases, the substratum had such large irregularities that the floor slab which was applied to it was too thin in places and it was unable to withstand the loads placed on it. If the site situation (slope, excavation to the base of the foundations) was such that it necessitated deep back-filling beneath the floor slab, cracking damage was observed in the floor slab and this could be attributed to heavy settlement of the back-filling material, which had been inadequately compacted.

Points for consideration

– The necessary thickness of the floor slab can only be achieved across the whole surface if the substratum has been carefully levelled. If, in order to avoid softening of a cohesive substratum by rain, the floor slab and the strip foundations are to be concreted in one operation immediately after the site has been excavated by machine, frequently careful levelling of the substratum will be neglected during construction work, so the footings and the floor slab will not be of the required shape and thickness.

– In general, the substratum supports the floor slab. If the substratum consists of back-filling material, it must be compacted in order to exclude the danger of subsequent heavy settlement of the back-filling material and thus sinking and in the eventual cracking of the floor slab. In examples where back-filling is to a depth in excess of 300 mm, the site should be back-filled and the material compacted in layers.

– Cohesive soils which are hard to compact and materials which do not have constant volume (e.g. slag, which swells when it absorbs water) are not suitable as back-filling materials.

– If the substratum is unable to provide support, or if it has been disturbed by building work (site traffic), a new supporting base can be produced by replacing the soil.

– Sufficient compaction of the substratum is of great importance if, after application of a protective coating of concrete, it acts as a sealing skin which is resistant to water under pressure, since this is the only way in which even compaction of the sealing layers can be guaranteed.

– Coarse fillings used as a substratum can act as a layer which interrupts capillary action because of the large cavities enclosed in them and can thus prevent soil moisture from rising under capillary action. In addition, if constructed in the form of surface drainage (see A 1.2) exposure of the basement floor to seepage water can also be reduced.

Recommendations for the avoidance of defects

● Attention should be paid to ensuring that the substratum of the basement floor slab is level and well compacted.

● The most reliable way of ensuring that the substratum is level is to use a smoothing layer of lean concrete about 80 mm thick.

● Where the site to be filled is deep, non-cohesive back-filling material which can also be easily compacted when damp, e.g. gravel, should be introduced in layers about 300–400 mm thick.

● If no sealing layer is going to be used in the basement floor where it is exposed to soil moisture (because the basement is not to be used for any major purpose) a compacted coarse filling at least 150 mm thick made of porous debris should be used as a seepage layer which will interrupt capillary action (see A 1.2.2).

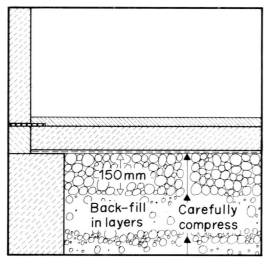

150 mm

Back-fill in layers — Carefully compress

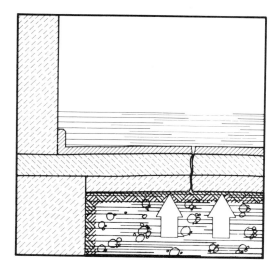

A number of examples of cracking damage were observed in the floor slabs of the basement floor. In most instances this damage was accompanied by extensive basement flooding, since in the majority of cases the floor was exposed (temporarily) to water under pressure.

The floor slab was either produced in one operation together with the strip foundations or was subsequently concreted on to the wall foundations. In none of the cases was the floor slab of suitable size to resist water under pressure nor was it in the form of a slab foundation.

Because of irregularities in the substratum, the thickness of the floor slab was irregular, and reinforcement was totally lacking or had been incorrectly inserted and had then been trodden down.

Points for consideration

– If the possibility, although only temporary, of the basement components being exposed to water under pressure is not taken into account, this has a particularly serious effect on the floor slabs if they are not part of the foundations. For whilst the foundation slabs are in the form of a rigid reinforced concrete slab in order to dissipate building loads across their surface and since the height of the water pressure has no considerable effect on the extent of the forces acting on the rigid floor slab (because soil pressure and the size of the lifting forces are reduced), floor slabs which are inserted between strip foundations and which are not of sufficient size to withstand surface loads from below are overloaded and will crack, even where they are only partially immersed in the groundwater.

– If a 60–80 mm thick concrete basement slab is designed, the floor slab will be so thin in some places because the substratum is not completely level that it will be unable to withstand the traffic loads and the slight settlement loads that occur. Cracks can then occur, for example when the footings settle more than the floor slab under the load of the building.

– If the strip foundations and the floor slab are to be concreted in one operation immediately after the site has been excavated by machine, the substratum is frequently not carefully levelled. The floor slab is then often not made to the correct thickness.

Recommendations for the avoidance of defects

● If careful determination of water pressure (see 0.1.1) has revealed that the basement floor can be expected to be exposed to water under pressure, the basement floor slab should take the form of a rigid reinforced concrete slab and its size should be determined by calculation.

● Floor slabs not exposed to water under pressure should be at least 100 mm thick. When applying concrete to coarse fillings or surface drainage a covering foil or polythene should be inserted in between them (for support of the floor slab on the foundations see C 2.1.2).

● Careful supervision should be used to check that the substratum is level, especially where the foundations and the floor slab are to be constructed in one operation.

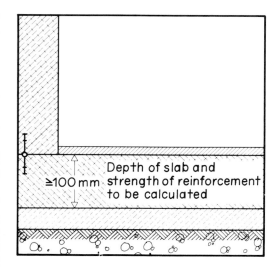

Depth of slab and
≧100 mm strength of reinforcement
to be calculated

Basement floor
Sequence of layers and individual layers

A large number of cases of damp damage were observed in basement floors where the 80–100 mm thick site-cast concrete floor slab was provided only with a binding screed and where the concrete was laid on a layer of material to interrupt capillary action.

Most of the basement floors were temporarily exposed to accumulated seepage water. Indeed, in a large number of cases, they were also exposed to water under pressure. No drainage measures to reduce the water load were provided, except in a few instances.

Frequently, the basement rooms concerned were to be used as living rooms (guest rooms, hobby rooms, bar) and were therefore provided with floating screeds and floor coverings which became damaged as a result of the penetrating damp.

Points for consideration

- The choice of a suitable method of moisture prevention depends both on the water pressure and on the purpose for which the basement is to be used.
- Increased demands are made in terms of the dryness of basement floors when they are used as living rooms (hobby rooms, guest rooms) and when they are to be used for the storage of goods sensitive to moisture. These increased demands must be met by suitable measures to provide protection against damp.
- Back-fillings which interrupt capillary action and cementitious screeds, which are the most economical type of conventional damp protection for the basement floor, may only reduce water seepage adequately enough to ensure that the floor surface is sufficiently dry where the pressure consists of soil moisture and where the basement is used for secondary purposes and where there is good ventilation.
- Although thoroughly mixed and well laid water-proof screeds and carefully applied sealing compounds can in individual cases prevent the passage of water even where there is exposure to water under pressure, the use of these sealing measures in the presence of water under pressure is very risky because of their liability to crack and because faults in their construction cannot be completely avoided. However, because of their position on the inside of the floor slab, faults which do occur are easily pinpointed and eliminated.
- Seals involving building papers which are applied to the floor slab offer good and reliable moisture protection, even where the floor is required to be very dry, but they are only suitable if applied in several layers and they can only be used in the case of exposure to water under pressure when an internal sump is provided to absorb the water pressure. Where this type of pressure is present, compressed sealing skins or tanking constructions produced by specialist firms are generally necessary beneath basement floors which are in the form of a slab foundation.

Recommendations for the avoidance of defects

● If heavy demands are made on the dryness of the basement floor, sealing layers should generally be applied to the floor slab, provided that possible exposure to water under pressure does not make it necessary to use tanking construction with building papers or one which is made from water-proof concrete.

● Layers which interrupt capillary action and cementitious screeds should only be used as the only means of providing protection against moisture where the pressure consists of soil moisture and where the basement is used for secondary purposes.

● Water-proof screeds and compounds should only be used when there is a guarantee that no seepage water can accumulate, for example by providing surface drainage, if necessary.

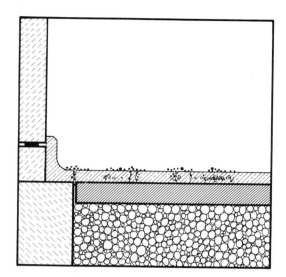

If basement floors were finished with a layer of cement screed as protection against moisture, damp penetration damage was frequently observed if the screed was supposed to act as a water-proof membrane, after the addition of sealing agents. In many cases, the screeds had cracked – the cracks either ran through the floor slab as well or affected only the screed. In most instances, the pressure consisted of accumulated seepage water or groundwater, so that there were not only damp patches and damage of floor finishes, but also puddles and flooding.

Points for consideration

– The moisture protection characteristics of bonded screeds depend on the dense composition of the screed material and on the freedom of the surface of the screed from cracks.

– The density of the screed material is largely produced by a texture which contains as few pores as possible. The grain size of the sand – and especially the proportion of fine grains – has a considerable influence on porosity. Similarly, a sufficiently large quantity of cement and minimum mixing water to ensure complete hydration are important. In mixing, however, it is important to ensure that the screed can be worked and compacted sufficiently. A designed mix and well compacted screed is therefore of greater importance to the moisture protection action of the screed than the addition of sealing agents.

– Additives can only increase the water-proofness of a screed that has already been well mixed and worked:

  – if they act as plasticisers, in other words if they improve the working properties of the screed (its ability to be compacted) whilst the water content remains low;

  – if they have the effect of constricting pores by swelling;

  – if they reduce the movement of water under capillary action by covering the walls of the pores with hydrophobic agents;

  – if they increase the bending strength of the screed and thus its resistance to cracking.

– As they are a relatively thin layer, screeds have low bending strength and are thus prone to cracking. Although cracks caused by inherent stresses in the screed (e.g. shrinking) and by loads imposed by use can be prevented by ensuring a solid bond with the substratum, the freedom of the screed from cracks caused by this bond is dependent on the freedom of the substratum from cracks.

– The strongest bond between the screed and the floor slab will be achieved if the screed is applied to the still wet floor surface one to two days after the floor slab has been made. However, further progress in the building work is then seriously impaired until the screed has fully hardened and the screed is exposed to a considerable risk of damage caused by subsequent work. If the screed is to be applied at a later date, the concrete substratum must be free of loose particles and must be slightly roughened in order to achieve a firm bond. Before applying the screed, this substratum must also be thoroughly wetted in order to ensure that the water the screed needs for hydration is not removed. Despite this, a cement-sand slurry brushed carefully into the substratum before the screed is applied will improve the keying of both layers.

– In the case of thin, non-reinforced floor slabs, there is no guarantee that the substratum will be free of cracks and, similarly, areas where different materials join and expansion joints that are to be covered by the screed are

Basement floor
Sequence of layers and individual layers

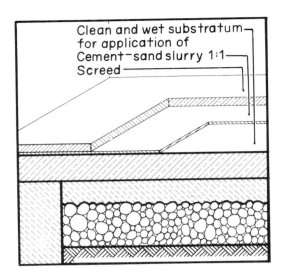

Clean and wet substratum
for application of
Cement-sand slurry 1:1
Screed

also likely to be areas where cracking occurs. Water-proof screeds therefore only produce a crack-free seal and a sufficiently strong protective layer on continuous, reinforced floor slabs.

– Screeds which have hardened or dried out unevenly are also prone to cracking. Screeded finishes must therefore be protected against excessively fast drying during setting and hardening and must if necessary be treated afterwards in an appropriate manner (kept damp).

Recommendations for the avoidance of defects

● Bonded screeds which are expected to provide protection against moisture should only be applied to level, reinforced, joint-less slabs.

● If subsequent building work permits, the screed should be applied to the still wet, clean floor slab one to two days after it has been laid.

● If the screed is to be applied to a floor slab which has already fully hardened the slab should be free of all impurities, loose particles and coarse irregularities and should be thoroughly wetted. A cement-sand slurry (mixing ratio about 1:1) should be brushed in immediately prior to applying the screed.

● Particular attention should be paid to ensuring that the composition of the cement screed is correct. The additive should consist of sand, grain size 0–3 mm with a proportion of finest grains of 0–0.25 mm of about 20 wt.%. The mixture ratio should be between 1:2 and 1:3 and the value of water in the cement should not exceed 0.5. Additives used according to the manufacturer's instructions may improve water-proofness.

● The screed should be applied in one operation to a thickness of about 30 mm and should be carefully compacted and smoothed without excessive rubbing and without the use of gauges.

● Until it has fully hardened, the screed should be protected against uneven and excessively fast drying (sometimes because of heat or currents of air and should not be used too soon.

Basement floor
Sequence of layers and individual layers

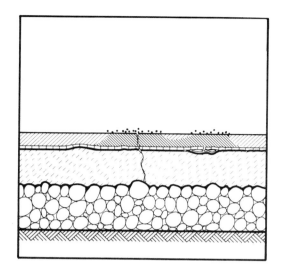

If sealing compounds were applied to the basement floor in order to provide protection against moisture, damp penetration nevertheless occurred which could generally be attributed to cracks and other flaws in the layer of compound which was just a few millimetres thick. The floors were exposed to accumulating seepage water which thus led to the formation of puddles.

Points for consideration

– As in the case of water-proof screeds, the water-proofness of sealing compounds depends not only on the dense structure of a fine cement-sand mixture with which pore constricting or water-proof additives are mixed, but also on the freedom of the layer of compound from cracks.

– Although faults in the optimum composition of the screed components can be largely excluded when the mixture is ready-made by the factory and when water is added according to the manufacturer's instructions, the danger of cracking and working faults when producing the thin, brittle layer of compound is very high. Similarly, the thin layer is easily damaged. In order to produce an effective sealing compound, therefore, it is of the utmost importance to ensure a good bond with the substratum, to ensure that the substratum has no flaws or cracks and that cracks or failures will not appear, to ensure that the necessary layer thickness is achieved and that the skin is permanently protected against mechanical damage.

– In order to achieve a firm bond with the substratum, the concrete surface must be clean, free of loose particles, level, but slightly textured, and it must be thoroughly wetted before the compound is applied in order to ensure that the water the setting layer needs to hydrate is not removed. Priming the substratum by applying a thin coating of compound also improves the key. The subsequent layers of compound must, however, be applied immediately after these priming compounds have set initially, otherwise the water-proof action of the hardened compound will have an adverse effect on the bond. For the same reason, the connection of the compound to parts of the structure which have already hardened also represents a defect.

– In the case of thin, non-reinforced floor slabs, or those which are made of low grade concrete, there is no guarantee that the substratum will be free of cracks, and similarly changes in material and joints which are to be covered by the compound also represent areas which are likely to crack. Thus, sealing compounds can only provide a crack free, and therefore water-proof layer on continuous, reinforced floor slabs.

– Because the whole of the compound layer is not very thick, even small irregularities can have a considerable effect on the seal. However, by applying the compound in several thin layers, each skin being applied after the previous one has begun to set, and by adhering to, or if necessary exceeding, the minimum coverage specified by the manufacturer, the risk of an uneven, excessively thin application can be avoided. If the compound is applied with a broom by workmen unfamiliar with this work uneven thicknesses will be produced. The compound should therefore be applied by skilled workers where possible using a float or a spraying device. The compound can only harden fully if the necessary water is removed from it, either by heat or by a current of air.

Basement floor
Sequence of layers and individual layers

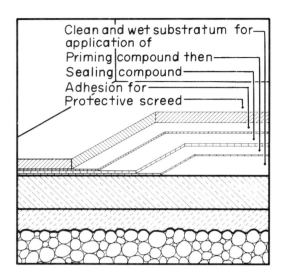

– Once it has hardened, the compound must be covered by a protective layer in order to protect it against mechanical damage. In examples of basement floors, this protection normally takes the form of a screed. In order to achieve a firm bond between the screed and the compound substratum an adhesion agent must be applied to the surface around the edges. When doing this, the compatibility of the materials should be taken into account.

Recommendations for the avoidance of defects

● Sealing compounds should only be applied to level, reinforced floor slabs which are free from flaws and joints.

● Before applying the sealing compound, the floor slab should be freed of impurities, loose particles and large irregularities, and should be thoroughly wetted. A priming compound should be brushed into the floor slab (according to the manufacturer's specifications). Once the priming compound has begun to set the sealing compound should be applied continuously in several layers. The minimum coverage specified by the manufacturer should be observed, and if necessary exceeded. The compound should be distributed and compacted with a float.

● Until it has fully hardened, the sealing compound should be protected against uneven and excessively fast drying out (to some extent by heat, current of air) and if necessary should be dampened.

● Immediately the compound has fully hardened, it should be protected against damage by applying a protective layer. This can take the form of a screed, whose bond with the compound is to be improved by applying an adhesion agent.

Basement floor
Sequence of layers and individual layers

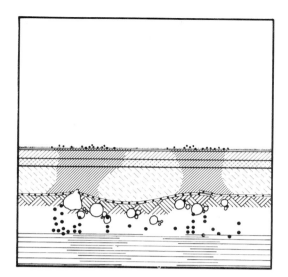

The majority of cases of damp damage to basement floors, such as saturation of the insulating layer of floating screeds or basement flooding, could be attributed to the failure of functional sealing layers. Damage to basement floors which were sealed with building papers or felts, however, was only found in special cases. Building papers or felts became damaged if they were laid on the base floor beneath the floor slab and if concrete was poured directly on them by mechanical means. Damage was also observed in single layer building papers which were applied without a protective layer to a floor slab which was not completely level. A single layer of grade 500 roofing felt was found to be unsuitable to withstand pressure from accumulated water.

Points for consideration

– Damage to building paper and felt seals in water-pressure resistant tanking constructions is relatively rare. This can be partly attributed to the fact that the specifications for this type of seal aim at a high degree of safety and to the fact that the designer and the builder are aware of the importance of a careful and fault-free contract, so that the sealing work is generally carried out by specialist firms.

– Leaks in the surface of building paper or felt seals against soil moisture and temporary exposure to accumulated water are usually the result of faults in the adhesion of the building paper courses, of damage sustained in building, and of movements in the substratum as well as of an excessively heavy water pressure being placed on the sealing layer.

– Although the risk of failures at the seams of the courses of building paper can be reduced by having wide overlaps of single layer building papers, a two layer covering with offset seams largely eliminates the possibility of faults. Since bituminous adhesive layers have an additional sealing function – in the case of non-coated films, this is indeed the only sealing function – the risk of flaws can be further reduced by applying adhesive layers evenly and fully and by applying a covering layer to these.

– The risk of damage caused by building work can be reduced by applying a protective layer immediately after laying. However, in order to ensure that any movements that might occur in the protective layer do not damage the building paper seal, a bond between the two layers must be prevented by means of a separating foil.

– Damage caused by distortion of the substratum can be avoided by ensuring that the bottom layer is sufficiently thick and uniform.

Recommendations for the avoidance of defects

● Building papers which are used as a seal against soil moisture and the temporary accumulation of seepage water should be applied to a clean, continuous floor slab at least 80 mm thick.

● Double skin seals with offset seams adhered to the whole surface and covered with a protective coating are preferable to single layer seals. Particular attention should be maintained in obtaining a sufficently wide overlap ($\geq$ 100 mm) and to ensuring that adhesion is correct.

● Immediately after laying, the building paper or felt seals should be provided with a protective covering (e.g. 50 mm of fine concrete, maximum grain size 3 mm) and this should be separated from the seal by a sheet of foil.

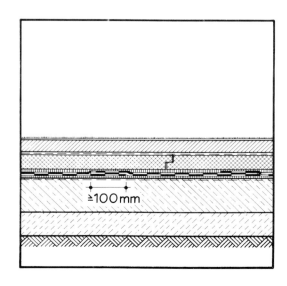

≥100mm

Basement floor
Sequence of layers and individual layers

In many instances, the water-proofness of normal concrete floor slabs is overestimated. Thus, in a number of cases, basement floors sustained damage where a floating screed had been applied directly to a concrete floor slab. It was assumed that a site-produced floor slab made of normal concrete and between 60 and 100 mm thick would be sufficiently resistant to soil moisture and even to accumulated water. The result of this was that the screed layers became damp, and sometimes cracks appeared which continued through all the layers. No damage was detected in reinforced water-proof concrete floor constructions.

Points for consideration

- The impermeability of a concrete depends on the density of its structure and on the freedom of its surface from flaws and cracks.

- The impermeability (lack of pores) of the structure depends mainly on the material and on the grain composition of the additives, on the cement content, on the quantity of water in the cement and on the way in which the concrete is poured. Only concretes which meet special requirements in answering these factors will produce a water-proof structure.

- Porous, water absorbent additives, as well as additives whose grain composition is irregular and does not correspond to the grading sequence, and an inadequate proportion of fine grain will produce a porous and thus permeable structure.

- The cement is the hydraulic binding agent of the concrete, and it forms part of the proportion of fine grain. Thus, in order to achieve a water-proof structure, a minimum proportion of cement must be used.

- Surplus mixing water which is not needed for hydration will result in pores in the concrete. The most favourable (lowest) proportion of water (expressed in the ratio of water to cement by weight, or water cement value) is not, however, the quantity of water needed for hydration (water cement value approx. 0.4) but also the quantity that will ensure the concrete is easily worked (plasticity) and compacted.

- Concrete additives can only improve the water-proofness of an already well mixed and worked concrete:

  - if they have a plastifying effect, i.e. if they improve the plasticity of cement containing a small proportion of water;
  - if they constrict the pores by means of components which swell;
  - if they prevent capillary water action by coating the pores with hydrophobic agents.

- Cracks can mainly appear in the floor slab as a result of varying degrees of settlement in the soil, as a result of distortion and longitudinal changes caused by inherent stresses and as a result of shrinking and creeping processes. Non-reinforced, or inadequately reinforced, floor slabs are particularly prone to cracking. If the floor slab is to provide protection against water under pressure as part of a water-proof concrete tanking it should be designed and reinforced in such a way as to limit cracking.

- Similarly, connecting joints and stage joints in the work also represent defects, and these should be avoided wherever possible, particularly in the case of water-proof concrete tanking constructions against water under pressure. If such joints are unavoidable, they should be designed carefully in advance.

Basement floor
Sequence of layers and individual layers

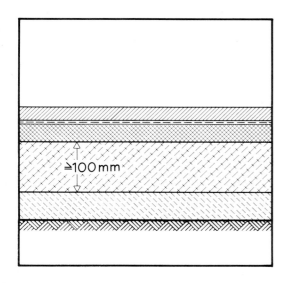

≥100mm

Recommendations for the avoidance of defects

● Water-proof concrete should be made from a dense additive with a maximum grain size of 32 mm with a regular grading sequence, plus at least 350 kg of cement per m³. The proportion of water in the cement must not exceed 0.6. Sealing agents can be added, but their suitability must have been proven. In the case of water-proof floor slabs which are to offer protection against water under pressure as part of a water-proof concrete tanking, the cross section (the reinforcement) should be of such a size that cracks are limited to a width of 0.1 mm.

● Joints in the work should be avoided wherever possible (if necessary by using retarding agents). If they are unavoidable, they should be carefully designed (e.g. by arranging jointing gaskets). The water-proof concrete should be introduced in such a way that separation is avoided and so that good compaction is achieved.

● Water-proof concrete floor slabs whose reinforcement does not take into account the limitation of cracking width, should only be used to provide protection against soil moisture and non-accumulating seepage water.

● In view of the large number of factors that have to be taken into account in order to achieve concrete which is water-proof when exposed to water under pressure, this work should only be carried out by specialist firms.

Basement floor
Sequence of layers and individual layers

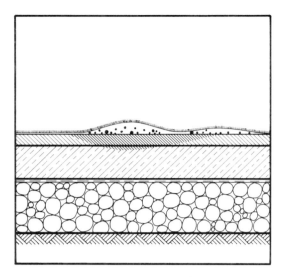

Patches of mildew on the basement floor, a musty smell and uncomfortable conditions inside the room were frequently observed in examples of living rooms whose floors were below the ground level, although they were provided with sufficient protection against moisture in the ground. In these instances, the basement floor slabs were frequently covered with a bonded screed and with a floor covering (carpet) – in other words, they had no thermal insulation layers.

The same forms of damage appeared in basement floors with floating screeds if their thermal insulation layer was not given sufficient protection against moisture, so that it became saturated, thus reducing its insulation effect.

Points for consideration

- In buildings which have no basement, as well as in living rooms in the basement (the latter are particularly common in residential buildings built on a slope) the basement floor must not only be protected against exposure to water in the soil but must also fulfil a thermal insulation function.

- At normal basement depths, the lowest soil temperature beneath the basement floor is approx. $+6°C$. The basement floor is thus not subjected to any major temperature fluctuations. The considerable heat losses are therefore not the result of low temperatures that occur periodically but of a small, but constant temperature drop.

- Similarly, despite the lack of extremely low temperatures, the formation of slight quantities of water as a result of prolonged condensation in basement floors which have inadequate thermal insulation will lead to damage, since the drying out periods which parts of the structure above ground level enjoy during the summer do not occur. In addition, the soil already has a higher moisture content than the air and this means that most sealing layers will be unable to withstand the constant reserves of moisture.

- Since the materials which are normally used in basement floors – in situ concrete, cement – have a negligible thermal insulation value which may be reduced depending on the type and position of the sealing layer, thermal protection must be provided by means of thermal insulation layers in order to reduce heat losses, to prevent condensation damage and to achieve a comfortable, warm floor surface in living rooms whose floors are below ground level.

- Recommendations for habitable basements therefore require floors which are below the ground to have a thermal transmission resistance of at least 0.86 m²K/W ($\hat{=}$ maximum transmission coefficient of 0.97 W/m²K) and a maximum thermal transmission coefficient of 0.90 W/m²K ($\hat{=}$ minimum thermal transmission resistance of 0.94 m²K/W).

- Since thermal insulation materials are absolutely essential in order to meet these requirements, the selection of a thicker insulation layer can produce increased thermal protection at slight additional cost. However, because of the fact that the winter temperature drop between the inside of the room and the ground is low, the energy saving achieved in this way is not as great as the saving made by equivalent improvements in the thermal insulation of structural components which are above ground level.

Recommendations for the avoidance of defects

● The floors of living rooms which are below the ground must have a thermal transmission resistance of at least 0.86 m²K/W ($\hat{=}$ maximum thermal transmission coefficient of 0.97 W/m²K) by using thermal insulation layers.

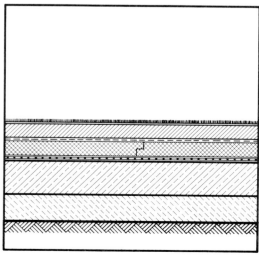

Basement floor
Sequence of layers and individual layers

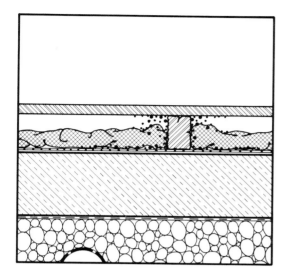

The thermal insulation value can be expected to be reduced if thermal insulation materials which are not moisture resistant or which are unable to absorb water are laid beneath the sealing layer. Condensation damage can occur if the air from the room has access to the cold side (underside) of the thermal insulation layer. This is easily possible with timber floors on joists.

Points for consideration

– The risk of damp penetrating thermal insulation layers must be avoided wherever possible, especially in structural components which are below ground level, since drying out is considerably less in these components than in those above ground level.

– Only closed-cell foam materials without open cavities between their particles (extruded polystyrene foams) have slight water absorption, even though they are constantly exposed to moisture. Thermal insulation layers made of this material therefore do not have to be protected against damp penetration by sealing layers and can be laid beneath the floor slab. However, current practice precludes at present the inclusion of insulation fitted beneath the sealing layer when assessing thermal protection.

– If thermal insulation materials applied on top of the sealing layer themselves have a low vapour barrier value (e.g. mineral fibre mats) and if no vapour-barrier layers are laid on top of the thermal insulation layers (e.g. ventilated suspended timber floors) condensation can occur throughout the year. In order to avoid damage, therefore, the upper surface of thermal insulation materials with a low vapour barrier value should be protected by a vapour barrier.

– Since most types of sealing layer are not completely water-proof (instead they merely slow down the percolation of water through the components to a greater or lesser extent) thermal insulation layers which may rot and which are not resistant to moisture will be damaged, particularly if they are covered from the inside by vapour-barriers. Similarly, absorbent thermal insulation materials will largely lose their thermal insulation value. If building paper seals are used, this danger of damp penetration can be prevented.

– Basement rooms with thermal insulation layers on the inside take less time to heat up, since the floor slab does not also have to be heated; they are thus of particular benefit only in the case of rooms which are used temporarily.

Recommendations for the avoidance of defects

● Only moisture resistant thermal insulation materials that cannot rot should be used to insulate the floors below ground level. As a rule, these materials should be applied above a seal of building paper.

● The upper surface of thermal insulation layers with a low vapour barrier value (e.g. mineral fibre mats) should be protected by a vapour barrier (e.g. laminated aluminium foil, 0.1 mm).

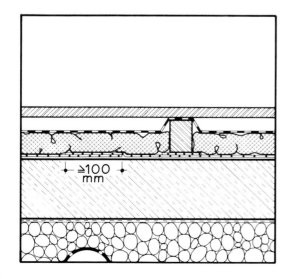

≥100 mm

Basement floor
Sequence of layers and individual layers

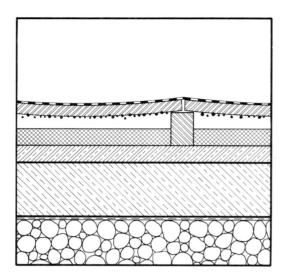

Damp damage to basement floors which were below the ground level was in a large number of instances reflected in damage to floor coverings. The thermal insulation layers of floating screeds swelled and warped and thus provided minimal thermal insulation; carpet coverings became rippled and showed signs of rotting; plastic coverings blistered; timber floors (chipboard, boards) showed signs of fungus and lost strength; parquet flooring warped.

The intention to use the basement as a living room was rendered impossible by this damage and the cost of renewing the damaged coverings was great.

Points for consideration

- Floor coverings which are sensitive to moisture require high quality sealing measures.

- Seals, with the exception of building paper seals or untearable felts, are not completely impermeable, even if they are laid correctly. They merely reduce the water action through the structural component to a greater or lesser extent. If relatively non-ventilated floor coverings (e.g. PVC coverings) are laid, diffusion of the water carried through the floor is prevented. This can then result in damp concentration beneath the floor covering (especially in the case of water-proof screeds or sealing compounds with protective screeds which have relatively high inherent dampness) and this reduces the thermal insulation value of the insulating layers of floating screeds and leads to losses in composition in the case of chipboard. In addition, damp concentration will promote fungal growth in all timber materials.

- Although in examples of timber floors laid on joists 'ventilation' of the air space is generally recommended by means of narrow slits along the skirting board, effective removal of moisture cannot be expected, however, because of this small cross section and because of the lack of currents to promote air movement.

Recommendations for the avoidance of defects

● If moisture-sensitive floor coverings (e.g. carpets, parquet flooring) are to be applied to floors below ground level, high quality sealing measures (e.g. building paper seals) should be used, even in the presence of slight exposure to water (soil moisture).

● Non-ventilated floor coverings (e.g. PVC coverings) should not be used on moisture-sensitive bases, or bases which might rot (e.g. chipboard) and particularly not on seals made of water-proof screeds and sealing compounds.

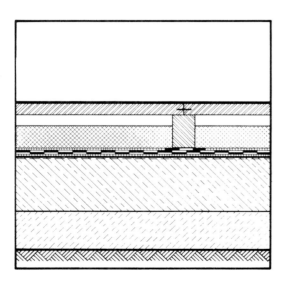

General specialist literature

Grassnick, Arno; Holzapfel, Walter: Der schadensfreie Hochbau, Grundlagen zur Vermeidung von Bauschäden. Verlagsgesellschaft Rudolf Müller, Köln 1976.

Lufsky, Karl: Bauwerksabdichtung, Bitumen und Kunststoffe in der Abdichtungstechnik. 3. Auflage, B. G. Teubner, Stuttgart 1975.

Reichert, Hubert: Abdichtungsmaßnahmen an erdberührten Bauteilen im Wohnungsbau. Forum Fortbildung Bau, Forum-Verlag, Stuttgart 1977, Heft 8, Seite 101–114.

Reichert, Hubert: Sperrschicht und Dichtschicht im Hochbau. Verlagsgesellschaft Rudolf Müller, Köln 1974.

Schild, Erich: Untersuchung der Bauschäden an Kellern, Dränagen und Gründungen. Forum Fortbildung Bau, Forum-Verlag, Stuttgart 1977, Heft 8, Seite 49–67.

AIB: Anweisung für die Abdichtung von Ingenieurbauwerken. Deutsche Bundesbahn, 2. Ausgabe 1953.

DIN 4031: Wasserdruckhaltende bituminöse Abdichtungen für Bauwerke, Richtlinien für Bemessung und Ausführung. November 1959.

DIN 4117: Abdichtung von Bauwerken gegen Bodenfeuchtigkeit, Richtlinien für die Ausführung. November 1960.

DIN 18195 – Teil 2 (Entwurf): Bauwerksabdichtungen, Stoffe. November 1977.

DIN 18195 – Teil 4 (Entwurf): Bauwerksabdichtungen, Abdichtungen gegen Bodenfeuchtigkeit, Ausführung und Bemessung. November 1977.

DIN 18195 – Teil 6 (Entwurf): Bauwerksabdichtungen, Abdichtungen gegen von außen drückendes Wasser, Ausführung und Bemessung. November 1977.

DIN 4122: Abdichtung von Bauwerken gegen nichtdrückendes Oberflächenwasser und Sickerwasser mit bituminösen Stoffen, Metallbändern und Kunststoffolien, Richtlinien. Juli 1968.

DIN 18336 – VOB, Teil C: Abdichtungen gegen drückendes Wasser. Oktober 1965.

DIN 18337 – VOB, Teil C: Abdichtung gegen nichtdrückendes Wasser. Februar 1961.

Floor slab

DIN 1054: Baugrund, Zulässige Belastung des Baugrundes. November 1969.

DIN 4019 – Blatt 1: Baugrund, Setzungsberechnungen. September 1974.

Sealing with a bonded screed

Albrecht, Walter; Mannherz, Ursula: Zusatzmittel, Anstrichstoffe, Hilfsmittel für Beton und Mörtel. 8. Auflage, Bauverlag, Wiesbaden und Berlin 1968.

Albrecht, Walter: Eigenschaften von Estrichmörteln und schwimmenden Estrichen. In: Boden, Wand + Decke, Heft 9/1966, Seite 742–760 und Heft 10/1966, Seite 870–889.

Piepenburg, Werner: Mörtel, Mauerwerk, Putz. 6. Auflage, Bauverlag, Wiesbaden und Berlin 1970.

Schütze, Wilhelm: Estrichmängel – Entstehen, Vermeiden, Beseitigen. Band 2, Industrie-Fußböden, Bauverlag, Wiesbaden und Berlin 1973.

Schütze, Wilhelm: Estriche und Abdichtungen gegen nichtdrückendes Wasser. In: Boden, Wand + Decke, Heft 2/1965, Seite 108–110.

Bundesverband Estriche und Beläge: Technische Richtlinien für die Verlegung schwimmender und im Verbund hergestellter Estriche. Bonn 1976.

DIN 18353 – VOB, Teil C: Estricharbeiten. 1973.

Sealing with sealing compounds

Brand, Hermann: Die zementgebundenen Oberflächendichtungsmittel. In: Das Baugewerbe, Heft 10/1976, Seite 35–36.

Grunau, Edvard B.: Bauwerksabdichtungen – Möglichkeiten und Methoden. In: Baugewerbe, Heft 18/1976, Seite 26–36.

Horstschäfer, Heinz-Josef: Nachträgliche Abdichtungen mit starren Innendichtungen. In: Forum Fortbildung Bau, Forum-Verlag, Stuttgart 1977, Heft 8, Seite 82–86.

Köneke, Rolf: Bauwerksabdichtungen nach DIN 4117 und DIN 18550 oder mit Zementschlämmen oder womit? In: Das Baugewerbe, Heft 20/1975, Seite 36–40.

Köster, Johann J.: Flächenabdichtungen bei Bauwerken. In: Das Baugewerbe, Heft 9/1975, Seite 31–32.

Schild, Erich: Nachbesserungsmaßnahmen bei Feuchtigkeitsschäden an Bauteilen im Erdreich. In: Forum Fortbildung Bau, Forum-Verlag, Stuttgart 1977, Heft 8, Seite 76–81.

Schumann, Dieter: Untersuchung über die Wirksamkeit von Dichtungsschlämmen; Abschlußbericht zu einem Forschungsauftrag des Bundesministers für Raumordnung, Bauwesen und Städtebau. Lehrstuhl für Baustoffkunde und Werkstoffprüfung der Technischen Universität München, München 1977.

Sealing with building papers

Braun, E.; Metelmann P.; Thun, D.: Bituminöse Hautdichtungen – Folgerungen aus Theorie und Praxis. In: Bitumen, Heft 5/1973, Seite 117–129.

Hauptverband der Deutschen Bauindustrie, Bundesfachabteilung Bauwerksabdichtung: Technische Regeln für die Planung und Ausführung von dehnfähigen Bauwerksabdichtungen. Otto Elsner-Verlagsgesellschaft, Darmstadt 1974.

Sealing with water-proof concrete

Albrecht, Walter; Mannherz, Ursula: Zusatzmittel, Anstrichstoffe, Hilfsstoffe für Beton und Mörtel. 8. Auflage, Bauverlag, Wiesbaden und Berlin 1968.

Albrecht, Walter: Über die Wirkung von Betondichtungsmitteln. In: Betonstein-Zeitung, Heft 10/1966, Seite 568–573.

Breuckmann, K.: Herstellen von Beton mit Betonzusatzmitteln. In: Baugewerbe, Heft 15/1976, Seite 14–17.

Gunau, Günter; Klawa, Norbert: Empfehlungen zur Fugengestaltung im unterirdischen Bauen. In: Die Bautechnik. Heft 10/1973, Seite 325–332.

Gundermann, Erich: Bautenschutz, Chemie und Technologie. 2. Auflage, Verlag Theodor Steinkopf, Dresden 1970.

Leonhardt, Fritz: Über die Kunst des Bewehrens von Stahlbetontragwerken. In: Beton- und Stahlbetonbau, Heft 8/1965, Seite 181–192.

Rapp, Günter: Technik des Sichtbetons. Beton-Verlag, Düsseldorf 1969.

Wesche, Karlhans: Baustoffe für tragende Bauteile, Band 2: nichtmetallisch-anorganische Stoffe: Beton und Mauerwerk: Bauverlag, Wiesbaden und Berlin 1974.

Wischers, G.: Zur Wirksamkeit von Betondichtungsmitteln. In: beton, Heft 8/1975.

DIN 1045: Beton- und Stahlbetonbau, Bemessung und Ausführung. Januar 1972.

Thermal insulation

Hamm, Paul: Perimeter – Dämmung = Kellerdämmung außenliegend. In: Deutsches Architektenblatt, Heft 2/1977, Seite 128–129.

Wallmeier, Jörg-Rüdiger: Nutzungsmöglichkeiten von Räumen mit erdberührten Außenwänden. Dissertation an der Fakultät für Bauwesen der RWTH Aachen 1974.

DIN 4108: Wärmeschutz im Hochbau. August 1969. Ergänzende Bestimmungen. Oktober 1974.

Wärmeschutzverordnung 1977. In: Bundesgesetzblatt, Teil 1, 1977, Seite 1554ff. und in: Deutsches Architektenblatt, Heft 9/1977, Seite 717–720.

Floor coverings

Rosenbaum, Erich; Burger, Hans; Bekic, Veljkor: Schadensfreie Fußböden – Planung, Ausschreibung und Verarbeitungstechnik. Bauverlag, Wiesbaden und Berlin 1971.

Zimmermann, Günther: Elastischer Sportboden in Mehrzweckhalle, Quellung und Festigkeitsminderung der Holzspanplatten. In: Deutsches Architektenblatt, Heft 4/1977.

Problem: Connection of basement floor to external and internal walls

In the damp protection system of the structural basement components, the connection of the floor to the external walls is of particular significance. A continuous, sealed or lapped connection must be made between the wall damp-proof course and the floor seal in an area which may be particularly heavily exposed to groundwater or to accumulated seepage water in the excavated site and in an area where the wall, floor and foundation (which are frequently made of different materials) connect.

Although in tanking construction which provides protection against water under pressure great attention is paid to this problem, detailed design and execution of this area is often neglected in damp-proofing against lesser water pressures. However, as these forms of damp-proofing are often applied in conjunction with drainage methods, and since the failure or temporary overloading of the drainage makes itself felt first and most strongly in the area of the base of the foundations, damp damage is particularly common in this context. Moreover, the connection is impaired in the case of sealing measures involving a damp layer on the inside of the floor slab and on the outside of the basement walls, since the floor damp-proof layer must be carried through the wall in cross section. For constructional reasons, the section of damp-proof course in the wall's cross section has to be inserted at an earlier date separately from the other floor and wall seals.

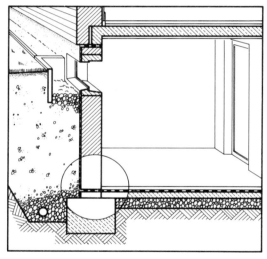

Faults in the connecting structure also arise from the fact that the designers and builders regard the horizontal damp-proof course laid in the wall cross section as a separate sealing measure providing protection against water rising under capillary action and therefore do not connect it to the wall and floor seal.

The unusually frequent damp damage which occurred as a result of leaks in the area where the basement floor connected with the external basement wall was generally restricted to the lower area of the wall, but complete damp penetration and basement flooding were also observed.

For the reasons mentioned, therefore, the defects discussed on the following pages concentrate on connecting structures in examples of sealing measures against exposure to soil moisture and against non-accumulating seepage water, or seepage water which accumulates only temporarily (instead of attempting to deal with the treatment of all possible material combinations), so that the most important aspects can be illustrated. The recommendations which are made are based on the fact that, in the case of damp-proof courses in the wall cross section, the great importance of and the extraordinary difficulties associated with remedial work require a high degree of reliability in the initial construction.

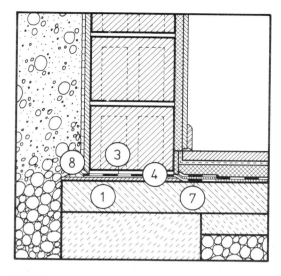

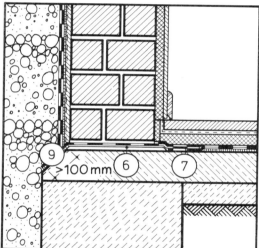

1 The basement floor slab should be seated directly on the strip foundations. If the building specification permits, and particularly if water-proof screeds and water-proof compounds are to be used as basement floor seals, the floor slab should be concreted as far as the outer edge of the foundations (see C 2.1.2).

2 In the case of water-proof concrete tanking, the connecting joint between the basement floor and the basement wall must be sealed by jointing gaskets inserted into the concrete, by pre-treating the adhesion surface and, if necessary, by means of a filling compound (see C 2.1.2).

3 If the basement sealing system does not consist of tanking construction a perfect seal should be made in the connection between the wall and the floor by means of a horizontal sealing layer in the wall (see C 2.1.3, C 2.1.5 and C 2.1.6).

4 Arrangement of the lower horizontal wall damp-proof course on a level with the basement floor seal is preferable, particularly if a design which is 100–150 mm higher considerably impedes the connection between the floor and the vertical wall seal (see C 2.1.3).

5 Heavy grade plastic films (Visqueen) or building papers with fibre inserts or lead core should be used as sealing material, especially when the path taken by the sealing course or the connections necessitates bends in the sealing layer (see C 2.1.4).

6 The contact surface for the sealing layer should be carefully levelled with mortar (mortar group III), the courses should be laid without adhesive with an overlap at the seams of at least 100 mm (in the case of exposure to accumulated water the seams should be glued or welded) and the work carried out should be inspected with care (see C 2.1.4).

7 The horizontal barrier layer in the wall made of reinforced building paper or heavy duty plastic film should be laid so that it projects by at least 100 mm. Basement floor seals made of water-proof screeds or sealing compounds should overlap this projection, and floor seals consisting of building paper or plastic films should be firmly glued to the projecting section (see C 2.1.5).

8 Where vertical seals made of bituminous coatings, water-proof render or sealing compound are applied to the external basement walls, the horizontal wall seal should be continued to the outside of the wall, including the render or compound layer. At this point, the horizontal seal should be cut off flush (see C 2.1.6).

9 Where the external basement walls are sealed with building papers, the horizontal wall seal should be laid so that it projects at least 100 mm. At this point, it should be firmly glued to the vertical wall seal (see C 2.1.6).

10 Internal basement walls should be included in the sealing methods for the base of the basement without voids. If this is not already guaranteed by a sealed foundation slab, a horizontal sealing layer should be laid to the internal basement wall. This seal should be made in the same way as the horizontal seal for external basement walls in terms of material selection, position and connecting structure (see C 2.1.7).

Basement floor
Connection of basement floor to external and internal walls

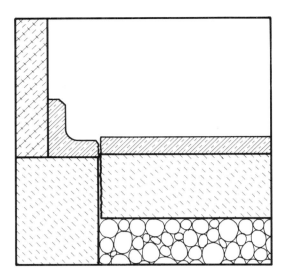

A number of instances of cracking and damp penetration damage stemmed from incorrect connecting and abutment structures of the basement floor slab where it joined the wall and the foundations.

If the basement floor slabs were not continuous on strip foundations but were concreted in between the latter, cracks appeared in the water-proof screed along the concrete joint and moisture was able to penetrate these cracks. Similar damage was observed in the connecting joints to the basement wall brickwork in examples of seals made of sealing compound and water-proof screeds if the floor slab was laid on the projection of the foundations.

If the basement wall was laid on a floor slab that took the form of a foundation slab, the floor and the wall had to act as a water-proof concrete tanking resistant to water under pressure and so damp also penetrated the concrete joints between the floor slab and the wall.

Points for consideration

– In most methods of sealing system the floor slab supports the sealing layer. If the floor slab is laid in such a way that it can move in relation to the wall (e.g. settlement) an increased risk of cracking can be expected around the connection of the floor seal to the wall seal – an area which is already prone to damage. This concerns brittle seals made of water-proof screeds or sealing compounds in particular. When applying these types of seal, therefore, it is necessary to have a level substratum which is firmly connected to the wall as, for example, in the case of a floor slab which is concreted over the foundations.

– In the case of water-proof concrete tanking constructions which are resistant to water under pressure, the joint where the wall is laid on the floor normally represents a working joint in which the two concrete surfaces produced at different times have to be connected with one another in a water-tight manner. Since the subsequently concreted wall section still has to undergo full shrinkage (although the floor slab has already undergone part of its own shrinkage) and since adhesion between the old and the new concrete is reduced (particularly in the case of concrete containing hydrophobic sealing additives) a crack will appear along the joint. Leaks can be avoided by inserting jointing gaskets in the concrete or by treating the adhesion surface in the appropriate manner (e.g. cleaning, roughening, or applying adhesion agents) and by filling the connecting joint. However, this type of complicated joint can generally only be made with sufficient care by specialist firms.

Recommendations for the avoidance of defects

● The basement floor slab should be seated directly on the strip foundations. If the building specification permits (floor concreted before the walls are built), and particularly if water-proof screeds and water-proof compounds are to be used as basement floor seals, the floor slab should be concreted as far as the outer edge of the foundations.

● In the case of water-proof concrete tanking, the connecting joint between the basement floor and the basement wall must be sealed by jointing gaskets inserted into the concrete, by pre-treating the adhesion surface and, if necessary, by means of a filling compound (see B 1.1.17).

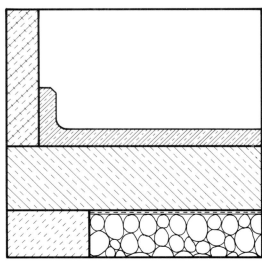

Basement floor
Connection of basement floor to external and internal walls

Damp damage in the area between the lower horizontal wall damp-proof course and the basement floor represents by far the most common type of damage in the basement area.

The basement walls generally consisted of brickwork, were coated externally with bituminous materials and had a horizontal wall damp-proof course, which usually took the form of a single layer of untearable felt about two courses of bricks above the floor slab. The edges of the damp protection layers – normally cementitious screeds – were not taken up sufficiently (or were not raised at all) to make contact with the lower horizontal wall seal. The area of wall below the damp-proof course showed signs of damp penetration and this damage also affected the edge of the floor coverings. If the lower horizontal wall damp-proof course was plastered over on the inside, the damp penetration in some cases extended far beyond this sealing layer.

Points for consideration

– The lower horizontal wall damp-proof course is not only intended to prevent moisture from rising up the wall under capillary action, instead it represents the connecting element in the basement wall between the floor seal and the vertical basement wall seal and is always necessary if the floor seal is applied to the inside whilst the wall seal is arranged on the outside of the floor or wall structures. If such a connection is not made, moisture is able to enter the basement through the connection between the wall and the floor.
– If the lower horizontal wall seal is arranged at a height of 100–150 mm above the top edge of the basement floor, the part of the basement wall above the sealing layer should be protected against moisture rising under capillary action and against any water standing on the bottom of the basement during construction. The connection of the floor seal is made more difficult because of the resultant difference in height, because of the deviated course it takes and because of the change in the substratum for the sealing layer, especially in examples of floor seals made of building papers and sealing compounds. Moreover, the appearance of the lower area of the wall inside the basement room is not good, even if the work is carried out correctly. For these reasons, therefore, it is common to find that no connection is made between the floor seal and the lower horizontal wall damp-proof course and that the sealing layer is plastered over after having been cut off flush with the wall surface.
– A horizontal wall seal arranged on a level with the sealing layer for the basement floor does not give rise to these difficulties and also offers protection against standing surface water in cases where the floor slab is not concreted until the basement has been completed. The possibly heavier moisture pressure and the risk of damage to the projecting edges of the sealing layer can be taken into account by appropriate selection of materials (see C 2.1.4).

Recommendations for the avoidance of defects

● If the basement sealing system does not consist of tanking construction a perfect seal should be made in the connection between the wall and the floor by means of a horizontal sealing layer in the wall.

● Arrangement of the lower horizontal wall damp-proof course on a level with the basement floor seal is preferable, particularly if a design which is 100–150 mm higher considerably impedes the connection between the floor and the vertical wall seal.

Basement floor
Connection of basement floor to external and internal walls

In lower horizontal wall damp-proof courses generally made of untearable felt, it was frequently noticed that water entered the building on a level with the sealing layer. These cases concerned mainly sealing layers which were applied to a badly prepared substratum a few courses of bricks above the basement floor as well as sealing layers applied on the same level as the basement floor. In most of the cases the lower horizontal wall damp-proof seal was not connected to the floor seal (generally a screed) and the connection to the vertical wall seal (generally bituminous coatings on render, water-proof rendering or sealing compounds) appeared to be cracked and was therefore not water-proof (especially where it was exposed to a water pressure above and beyond soil moisture).

In some instances it was noticed that the wall had moved in the area of the lower horizontal wall seal. In these cases, the excavated site had been back-filled shortly after the wall had been built.

Points for consideration

– When selecting the material for the lower horizontal wall damp-proof course and when constructing this seal the following aspects should be taken into account: permanent effectiveness, a water-proof connection to the vertical wall seal and to the floor seal, and to the absorption of horizontal loads (pressure of the soil, water pressure) by the structure of the basement wall.

– As a horizontal layer which is inserted into the supporting walls, the damp-proof course is exposed to the full load of the building. For this reason, no sealing materials should be used which will lose their effectiveness under a load, such as, for example, those which will move to the side. In the case of thin sealing layers even small irregularities in the contact surface can result in failure.

– Since even smooth sealing layers have a high friction resistance once the full load of the building is applied to them, and because of the stabilising effect of external and internal walls running at right angles to the wall surface, the danger of a lateral move on a level with the inserted seal (which for example gives the impression that it is necessary to apply adhesive to the whole surface of the sealing layer) only normally arises when the basement wall is erected. However, firm connection (adhesion, welding) of the seams of the damp-proof courses is necessary where there is increased exposure to water.

– Where sealing layers are applied dry, stricter requirements are made in terms of achieving contact of the whole surface of the sealing layer without any voids, otherwise free water will be able to enter the room through the dry joint on a level with the sealing layer.

– Faults in the horizontal sealing layer in the wall cannot be repaired in a satisfactory manner after building has been completed – or at least this can only be done at extremely high cost. The material selected for the horizontal wall seal and the way in which the seal is constructed should thus aim at minimising the risk of failure. It is considerably easier to recognise and check faults and flaws in seals made of bituminous sealing papers and plastic films than in seals made of layers of water-proof rendering or sealing compound.

– Basically, the material for the lower horizontal wall damp-proof course must be selected to make it possible to produce a continuous seal between the sealing layers on the floor and on the wall in a simple manner. However, as the sealing

Basement floor
Connection of basement floor to external and internal walls

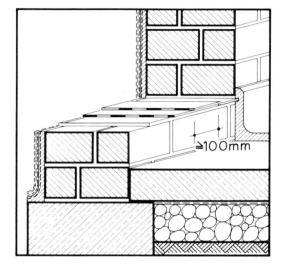

layer in the wall was made earlier than the other sealing layers in the basement area, the correct connection of layers of water-proof rendering and sealing compound layers to a lower horizontal wall seal of the same material necessitates special measures (because of the water-proof effect of the layers once they have hardened) which are not much easier than connecting them to inlaid building papers.

Recommendations for the avoidance of defects

● Heavy grade plastic films (Visqueen) or building papers with fibre inserts or lead core should be used as sealing material, especially when the path of the sealing course or the connections necessitate bends in the sealing layer, or if the edges of the sealing course can be expected to suffer heavy mechanical loads during construction.

● The contact surface for the sealing layer should be carefully levelled with mortar (mortar group III), the courses should be laid without adhesive with an overlap at the seams of at least 100 mm and the work carried out should be inspected with care.

● In examples of reinforced building paper seals against accumulating seepage water, the seams of the horizontal wall seal, which should be at least 100 mm wide, should be adhered or welded so they are water-proof.

Basement floor
Connection of basement floor to external and internal walls

In the case of lower horizontal wall damp-proof courses (normally untearable felt) arranged 100–150 mm above the basement floor, damp penetration was observed below the level of the seal in a large number of instances, even though the vertical external seal (generally bituminous coverings) was made correctly and was connected to the horizontal sealing layer. In all these instances there was no connection of the floor seal to the lower horizontal wall damp-proof seal – the floor sealing layer (generally a screed) frequently ended flush with the external basement wall. If the lower horizontal wall damp-proof seal did not once reach as far as the internal wall surface, connection was in no way possible.

A smaller number of instances of damp penetration along the floor connecting joint also occurred in horizontal seals at the same height as the floor seal; in these cases, the connection between the two seals was not water-proof.

Points for consideration

– Even if the external vertical wall seal is continued as far as the foundation where it is connected with a cement fillet, the section of wall beneath a horizontal damp-proof seal fitted 100–150 mm above the floor slab can become damp, since it is in contact with the foundation, which is not protected against moisture, and with the floor slab. If this section of wall is not covered on the inside by the floor seal which is continued as far as the lower horizontal seal, this damp damage will be visible on the inside of the room.

– Where horizontal wall damp-proof seals are fitted at the same height as the basement floor there is no problem of areas of wall to be sealed internally (see C 2.1.3). Since basement seals which necessitate a lower horizontal wall damp-proof seal are generally only used in instances of exposure to soil moisture (or to seepage water which accumulates for only short times), the production of a sufficiently water-proof connection between the floor seal and the horizontal wall seal can be achieved at no great expense. In exposure to soil dampness, the appearance of a moisture bridge must be prevented by overlapping the two sealing layers sufficiently. Where there is exposure to seepage water which accumulates for short periods, this overlap must be sealed by adhesive. In order to achieve such an overlap, the horizontal wall seal made of reinforced building paper or heavy duty plastic film must generally project at least 100 mm from the wall surface. Where horizontal wall seals are fitted on the same level as the floor slab, this type of overlap is possible at low cost, even in examples of connection to water-proof screeds and sealing compounds.

Recommendations for the avoidance of defects

● The lower horizontal wall damp-proof course must be connected to the sealing layer of the basement floor.

● The horizontal barrier layer in the wall (on a level with the surface of the basement floor wherever possible) made of reinforced building paper or heavy duty plastic film should be laid so that it projects by at least 100 mm. Basement floor seals made of water-proof screeds or sealing compounds should overlap this projection, and floor seals consisting of building paper or plastic films should be firmly adhered to the projecting section.

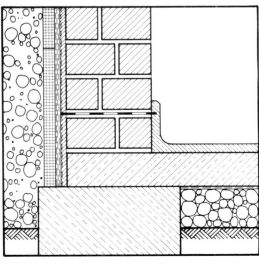

Basement floor
Connection of basement floor to external and internal walls

Above the lower horizontal wall damp-proof seal, basement walls often had a damp area, the height of which was mainly dependent on the extent of the external water pressure. The following faults were observed at the point where the lower horizontal wall damp-proof seal was connected to the vertical external wall damp-proof seal.

Although vertical wall seals made of bituminous coatings on render were continued as far as the foundations and indeed were frequently connected to the foundations by means of a cement fillet, the horizontal wall damp-proof seal made of untearable felt, however, ended inside the wall or beneath the render and so connection of the two sealing layers was impossible. If the floor and wall seals consisted of single-layer building papers, the two sealed surfaces were either merely butted flush together along their contact edges or were overlapped by only a few centimetres and were not carefully adhered to one another.

Points for consideration

– The protection which the lower horizontal wall damp-proof seal affords the section of wall above it is lost if no connection is made to the vertical wall or if this connection is inadequate. Greater attention should thus be paid to obtaining a firm connection – and not only in the case of water-pressure resistant tanking constructions.

– An overlap with the vertical building papers is only possible, and a sufficiently water-proof connection to coatings, water-proof rendering or sealing compounds can only be ensured, if the horizontal sealing layers inserted into the wall are wider than the cross section of the wall.

– Even if the vertical wall damp-proof seal is continued from above the horizontal wall damp-proof seal to the projection of the foundations, and if it is connected to the foundations by means of a cement fillet, the damp penetration of the area of wall beneath the horizontal seal as a result of water coming from the side can be reduced, but not eliminated, since this area of the wall can become damp through the foundations. However, this type of construction in no way obviates the need for a firm connection of the horizontal wall damp-proof seal to the vertical wall damp-proof seal and to the floor seal. If the horizontal wall seal runs down to the projection of the foundations, the cement fillet enlarges the contact surface between the two sealing layers to be connected and thus increases the water-proofness of the connection.

Recommendations for the avoidance of defects

● The lower horizontal wall damp-proof seal must be connected to the vertical damp-proof seal of the external basement wall.

● Where vertical seals made of bituminous coatings, water-proof render or sealing compound are applied to the external basement walls, the horizontal wall seal should be continued to the outside of the wall, including the render or compound layer. At this point, the horizontal seal should be cut off flush.

● Where the external basement walls are sealed with building papers, the horizontal wall seal should be laid so that it projects at least 100 mm. At this point, it should be firmly adhered to the vertical wall seal.

● As an additional measure, the vertical sealing layer can be continued from above the horizontal sealing layer to the projection of the foundations where it can be connected with a cement fillet.

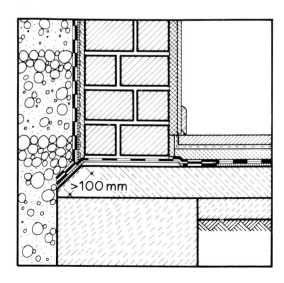

>100 mm

Basement floor
Connection of basement floor to external and internal walls

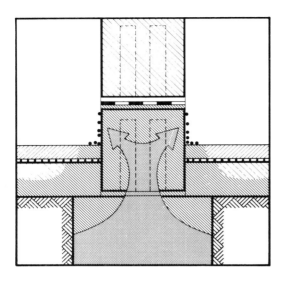

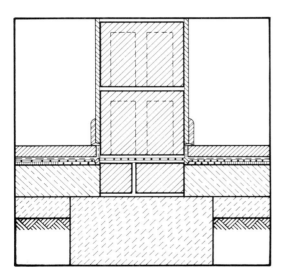

If the floor seal was not continuous through the internal walls in the form of a horizontal sealing layer, or if the floor sealing layer was not connected to the lower horizontal internal wall seal which was arranged about 100–150 mm above the basement floor, damp penetration was also observed on the internal walls.

If in this instance the lower horizontal damp-proof layer was also continued to the wall surface, damp penetration was limited to the area beneath the sealing layer. If, however, the edges of the seal had been plastered over, or if there was no sealing layer whatsoever, the damp area in some cases rose high up the wall.

Points for consideration

– If the internal basement walls are laid on strip foundations or on a continuous floor slab, moisture can enter the internal wall under capillary action through this bearing surface. Like the external basement walls, therefore, the internal basement walls must also be provided with a horizontal damp-proof layer which is continuously connected to the basement floor seal. This is unnecessary if the whole surface of a continuous floor slab is sealed before the internal walls are erected.

– The problems of arrangement, material selection and connecting structures for the lower horizontal wall damp-proof seal of the internal wall are largely the same as those for the horizontal seal of the external basement wall (see C 2.1.3, C 2.1.4, C 2.1.5).

Recommendations for the avoidance of defects

● Internal basement walls should be included in the sealing methods for the base of the basement without voids. If this is not already guaranteed by a sealed foundation slab, a horizontal sealing layer should be laid to the internal basement wall.

● This sealing layer should generally consist of reinforced building paper or heavy plastic films, to be fitted at the same level as the basement floor seal. The edges of the wall seal should be connected to the basement floor seal to form a water-proof joint (see C 2.1.5).

Specialist literature and directives

Beuße, Horst: Ursachen von Fußbodendurchfeuchtungen in nichtunter-
kellerten Räumen. In: Architekt + Ingenieur, Heft 1/1970, Seite
B 1–B 3.

Grassnick, Arno; Holzapfel, Walter: Der schadensfreie Hochbau. Verlagsgesellschaft Rudolf Müller, Köln 1976.

Lufsky, Karl: Bauwerksabdichtung, Bitumen und Kunststoffe in der
Abdichtungstechnik. 3. Auflage, B. G. Teubner, Stuttgart 1975.

Reichert, Hubert: Sperrschicht und Dichtschicht im Hochbau. Verlagsgesellschaft Rudolf Müller, Köln 1974.

Reichert, Hubert: Abdichtungsmaßnahmen an erdberührten Bauteilen im
Wohnungsbau. Forum Fortbildung Bau, Heft 8, Forum-Verlag, Stuttgart 1977, Seite 101–114.

Schild, Erich: Untersuchung der Bauschäden an Kellern, Dränagen und
Gründungen. Forum Fortbildung Bau, Heft 8, Forum-Verlag, Stuttgart
1977, Seite 76–81.

DIN 4117: Abdichtung von Bauwerken gegen Bodenfeuchtigkeit, Richtlinien für die Ausführung. November 1960.

DIN 18 195 – Teil 4 (Entwurf): Bauwerksabdichtungen, Abdichtungen gegen Bodenfeuchtigkeit. Ausführung und Bemessung. November 1977.

English language bibliography

Bowyer, J. and Trill, J. E. (1973) *Problems in building construction* Book 1 and Tutor's guide. Architectural Press.

British Standards Institution, London
Standards and Codes of Practice:
  BS 988, 1076, 1097, 1451: 1973 *Mastic asphalt for building (limestone aggregate)*.
  BS 1162, 1418, 1410: 1973 *Mastic asphalt for building (natural rock asphalt aggregate)*.
  BS 1194: 1969 *Concrete porous pipes for underground drainage*.
  BS 1186: 1971 *Clayware field drain pipes*.
  BS 3882: 1965 *Recommendations and classification for top soil*.
  BS 5642 Part 1: 1978 *Sills and copings*.
  CP 101: 1963 *Foundations and sub-structures for non-industrial buildings of not more than four storeys*.
  CP 102: 1973 *Protection of buildings against water from the ground*.
  CP 303: 1952 *Surface water and sub soil drainage*.
  CP 2001: 1957 *Site investigation*.

Building Research Establishment, Garston
Digest 54 *Damp-proofing solid floors*.
Digest 77 *Damp-proof courses*.
Digest 90 *Concrete in sulphate bearing clays and ground-water*.

Current papers, No. 4 (1978) S. A. Covington. *Basements in housing*.
Research Series, Vol. 3 (1978) *Foundations and soil technology*.

Construction Industry Research & Information Association (1978)
Guide to the design of water-proof basements.

Department of the Environment
Condensation in dwellings, Parts 1 and 2. London HMSO.
Advisory leaflet 51 *Watertight basements*, Part 1. London, HMSO.
Advisory leaflet 52 *Watertight basements*, Part 2. London, HMSO.

Duell, J. and Lawson, F. (1977) *Damp-proof course detailing*. Architectural Press.
Elder, A. J. and Vandenberg, M. (1974) *Architects Journal handbook of building enclosure*. Architectural Press.
Foster, J. S. and Harrington, R. (1973 & 1976) *Structure and Fabric*, Parts 1 and 2. Batsford.
Handyside, C. C. (1974) *Everyday details*. Architectural Press.
Harrison, D. (Ed) (1978) *Specification Vols 2 and 3*. Architectural Press.
Woolley, E. L. (1978) *Drainage Details in SI Metric*. Northwood Publications.

Index